Practicing Science, Living Faith

Columbia Series in Science and Religion

THE COLUMBIA SERIES IN SCIENCE AND RELIGION

The Columbia Series in Science and Religion is sponsored by the Center for the Study of Science and Religion (CSSR) at Columbia University. It is a forum for the examination of issues that lie at the boundary of these two complementary ways of comprehending the world and our place in it. By examining the intersections between one or more of the sciences and one or more religions, the CSSR hopes to stimulate dialogue and encourage understanding.

The Faith of Biology and the Biology of Faith
Robert Pollack

Buddhism and Science: Breaking New Ground
B. Alan Wallace, ed.

Environmental Ethics, Ecological Theory, and Natural Selection: Suffering and Responsibility
Lisa Sideris

William James and a Science of Religions: Reexperiencing The Varieties of Religious Experience
Wayne Proudfoot, ed.

Spirit, Mind, and Brain: A Psychoanalytic Examination of Spirituality and Religion
Mortimer Ostow

Contemplative Science: Where Buddhism and Neuroscience Converge
B. Alan Wallace

Practicing Science, Living Faith

INTERVIEWS WITH TWELVE LEADING SCIENTISTS

Edited by

Philip Clayton and
Jim Schaal

Foreword by William Phillips

Columbia University Press New York

Columbia University Press
Publishers Since 1893
New York Chichester, West Sussex

Library of Congress Cataloging-in-Publication Data

Practicing science, living faith :
interviews with twelve leading scientists /
edited by Philip Clayton and Jim Schaal ; foreword by William Phillips.
p. cm. — (The Columbia series in science and religion)
ISBN-10 0–231–13576–9 (cloth : alk. paper)
ISBN-13 978–0–231–13576–4 (cloth : alk. paper)
1. Religion and science. 2. Scientists—Religion.
3. Scientists—Interviews.
I. Clayton, Philip. II. Schaal, Jim. III. Title. IV. Series.

B:240.3.P73 2007
201' .65—dc22 2006023326

Columbia University Press books are printed on
permanent and durable acid-free paper.

This book is printed on paper with recycled content.

Printed in the United States of America

c 10 9 8 7 6 5 4 3 2 1

Contents

Foreword

William D. Phillips

When I was first invited to participate in the second phase of Science and the Spiritual Quest, I was of two minds. On the one hand, I could see the obvious benefits of discussing matters of spirit and faith with other physicists. I was increasingly being asked to speak about the relationship of science and religion, and I felt quite ill prepared. Interacting with intelligent and interested physicists of varied perspectives seemed like an obvious way to prepare myself. On the other hand, it's hard to put the topic of science and religion at the top of one's list of important issues while involved in a demanding program of scientific research. I wondered whether the SSQ commitment would be worth the time, travel, and effort it would require.

Despite my initial doubts, I plunged into the program and have never regretted it. The process began with my being interviewed by Philip Clayton, one of the SSQ principal investigators. Similar interviews started the SSQ experience for all the other participating scientists; twelve of the most memorable are gathered here. The interviews provided a way for many of us to begin thinking about the science-spirit intersection, as well as a means for introducing the various scientists to one another. These interviews then paved the way for the rest of SSQ: private meetings with discipline-specific workshops (with some cross-disciplinary discussions), and public events in which scientists, informed by the private discussions, shared their thoughts about science and the spiritual quest with general audiences.

As I had expected, the discussions with colleagues in the physics and cosmology workshops did indeed provide me with many important insights that have since allowed me to speak more coherently about science and faith. More important, I have come to a deeper understanding of how my science and my religious faith intersect and inform each other. And most important, I have made new and enduring friendships, relationships that are both academic and personal—the kinds of connections that can make all the difference in professional and private life. Robert Pollack speaks eloquently in his interview of how important it is for us to acknowledge and embrace the personal, emotional part of our lives as scientists and not just the objective, quantitative part. I consider it an important gift that a group of physicists, initially strangers to one another, was able to achieve the trust and friendship needed to discuss matters that are not normally part of academic exchanges.

One of the features that set SSQ2 apart from SSQ1 was that the scope of inquiry was broadened to include scientists with faith backgrounds beyond the Western traditions. Our panel included individuals whose roots were Jewish, Christian, Muslim, Buddhist, Hindu, and so forth, at various stages of faith and doubt, spanning the spectrum from devout to atheist. I was apprehensive about whether such a diverse group would have any chance of fruitful discussion about such a controversial and personal matter as religious faith and its relation to science. In fact, I was amazed that our common language and culture as physicists, along with an attitude of humility and mutual respect, easily bridged the other differences. This rapport made the SSQ experience unexpectedly fruitful for me. In reading the interviews that make up this book, I am again struck not by the differences but by the deep similarities that run through them like an underground river. One discerns here certain unmistakeable commonalities in the way that scientists approach the question of religious beliefs and values—even when they come from quite distinct disciplines and from vastly different cultures around the world.

The public events for which SSQ became famous brought their own revelations, no less important than those of the private discussions. Folks from all walks of life traveled great distances to be present at the public lectures that were being held around the world; they were eager both to listen to the lectures and to engage the speakers in conversation after the formal sessions. I believe that the overwhelmingly positive public response to scientists' talking about their personal beliefs on spiritual matters was in large part a result of the collegiality and personal warmth that developed between the participants. The public has frequently seen the science-religion question in terms of debate and conflict. Too often people are told that these two important areas of life are in fundamental opposition, with the defenders of each characterizing the other as apostates or ignoramuses. The thousands of attendees at public SSQ events saw none of that. Rather, they saw what the reader of this book sees: thoughtful men and women describing the different ways in which they handle the scientific and spiritual spheres of their life while respecting and valuing the experiences of those who have found different paths.

In studying the interviews I am again surprised to discover that so many core values of my own faith are shared by physicists of very different faiths. It is obvious, of course, that the great religions of the world have significant differences, especially at the level of specific metaphysical beliefs. Still, in this book dedicated to the lives of scientists who are also religious practitioners, one is repeatedly struck more by the similarities than by the differences.

It is also fascinating to observe the contrasts in the ways that scientists from different disciplines see their respective worlds. Compare the interviews with the biologists with those of the physicists, for example. For the most part, physicists find great beauty and harmony in the simplicity and symmetry of physical law. (The reader will see this feature of a physicist's view of nature shining through in the interview with Khalil Chamcham.) By contrast, the SSQ biologists whose work is featured here tend to see nature as messy and rather more random. Nonetheless, scientists from all the disciplines represented here clearly find deep spiritual connections between their practice of science and their faith.

I write this foreword as I return to the United States from the Vatican. There, together with an international group of scientists from vastly different disciplines, faiths, and cultures—in this sense much like the Science and the Spiritual Quest participants—I attended my first meeting of the Pontifical Academy of Sciences. Whether at the Pontifical Academy or SSQ, one cannot help but feel that, in a world increasingly filled with contention and brokenness, the friendships and personal relationships within the diverse community of science can act to bring healing. While it is only a small start, I hope that this book of spiritual reflections from scientists around the globe will help its readers to find a calm and valuable refuge from a tempest of conflict about science and spirit.

Preface

W. Mark Richardson

The late Thomas Odhiambo grew up in the spectacular beauty of the high plains of Kenya. The mountains, rivers, and trees were living characters in his village's collective story, and so too were the spirits of ancestors. No rendering of his community's practices and values could be given without reference to these elements. In his youth Odhiambo was sent to an Anglican school, where he absorbed the values and ideas of the post-Enlightenment West as if learning a second language. He then received a graduate education at the University of Cambridge and began what would be a distinguished career as an internationally recognized entomologist.

For the rest of his life Thomas Odhiambo seized upon the best of two cultures to motivate his work. With extensive knowledge of the tsetse fly, he developed technologies to solve major agricultural and health problems in Kenya, providing solutions that were appropriate in scale to African agriculture and sensitive to wider cultural and ecological effects. Here one sees the integration in a single life of scientific, political, moral, and aesthetic dimensions. When Thomas Odhiambo, a scientist steeped in a spiritual heritage spanning countless generations, died in 2003, he was president emeritus of the African Academy of Sciences, a sought-after speaker for UNESCO and other organizations, and a highly respected international figure.

This story offers but one illustration of the diverse group of twelve scientists represented in this book. Each interview provides a unique glimpse into the fundamental intertwining of scientific, moral, and spiritual influences in the careers of leading scientists. They allow us to see the connective tissues that bind these various aspects within a single human life.

The Science and the Spiritual Quest (SSQ) project, begun in 1996 at the Center for Theology and the Natural Sciences in Berkeley, California, was an experiment in dialogue at the intersection of these two domains. Between 1996 and 2003 SSQ provided a forum for more than 120 leading scientists from many disciplines—physics, astrophysics, biology, the neurosciences, information technology, and computer science—and from many different regions and religions around the world. Each scientist participated in intensive workshops with colleagues who were from the same or similar scientific fields but who represented a wide range of cultural and religious histories. The task of the participants was to probe and explore the connections they found between their shared experiences as scientists and the moral and religious commitments that also define their lives and work.

SSQ approached this nexus with a personal focus, believing that biography and narrative are irreplaceable genres for expressing the variability and contingency in the ways scientists fold together their moral and spiritual commitments and their professional practice. How is the ultimate personal meaning by which a scientist lives also informed by her scientific profession? In what ways do ethical commitments and beliefs about the ultimate transcend scientific judgment? No single formula is capable of answering such questions.

We who organized the project were impressed by the diversity among the kinds of connections the SSQ scientists wished to explore and by the goodwill, mutual respect, and humble spirit they brought to this inquiry. We realized that, in an era when core beliefs and values frequently become a source of division and violence, we were witnessing a radical alternative. Admittedly, interreligious dialogue does not always take place in the ideal conditions of the SSQ workshops in Paris, New York, Berkeley, and Stanford. But the depth and constructive spirit of these historic meetings among scientists also reflected the success of these scientists in finding a way to apply the hypothetical mode of thinking, already familiar from their scientific practice, as a standard for discussing the holistic commitments that characterize religious life and moral commitment.

The first iteration of SSQ, culminating in a widely covered international conference in Berkeley in 1998, was largely focused on cognitive connections of various kinds between scientific theories and those basic features of world interpretation that are expressed in the world's religious traditions. It was a mammoth task. Not surprisingly, the successes came not in well-developed systems but in more modest forms: identifying fruitful points for discussion, noting common ground with colleagues, laying out differences clearly and without defensiveness. In the scientists' debates one could discern four kinds of interest at play:

Speculation on the metaphysical and religious implications of cutting-edge work in science
Discovery of parallels between science and religion at the levels of practice, experience, and attitude
Pursuit of relationships, whether harmonious or tense, between the beliefs of one's religious tradition and the scientific theories with which one works
Exploration of the ethical implications of scientific research and practice, including not only the areas of science in which one specializes but also the ethics of one's application of science to human life generally

The second major project, SSQ2, from which this book mainly derives, further developed these high-level dialogues but now with a greater emphasis on non-Western and nontheistic religious traditions and interpreting *spirituality* in the broadest sense possible. Of course, the scientists' theoretical concerns run through these interviews as well. But perceptive readers will also hear overtones of the more immediate and practical concerns on which the SSQ program focused as it sought to bring science, ethics, and religious life into a single conversation.

One will quickly realize that no single theoretical model can capture the way scientists integrate their life's work with ethical or spiritual values—these conversations are too rich for simple formulas. Some scientists gradually discovered a deep values-based motivation that took hold of their scientific work and began to affect its subsequent development. Others entered their science already possessing a well-formulated framework of values that could guide the kinds of questions they asked and the applications they pursued. Still others described a dialectical relation between the apparent independence of science and values and the inescapability of the moral domain in human life and action.

A challenge lies just under the surface in these discussions: Can criteria be found for justifying results at the nexus of science and ethics? Despite the diverse histories from which values spring, are there nevertheless universal norms of reasoning that can guide the integration of human practices and resolve our differences? Or are human norms themselves so context dependent that each system of values remains opaque to the next, and differences in outlook remain curiosities to be observed rather than resolved? Does the role of ethics in the context of science ever get beyond personal preference to the sort of shared assumptions that make public discourse possible?

This book provides ample material for readers to draw conclusions of their own. Personally, when I read the story of Thomas Odhiambo, along with the others presented here, I cannot help but conclude that, despite the occasional strangeness of our respective worlds to each other, something fundamental about our shared humanity is expressed in these stories and the core commitments that bind these lives together.

The editors have done a superb job of conveying the heart and soul of Science and the Spiritual Quest. As organizers of the last four years of these summit meetings among scientists, Philip Clayton and Jim Schaal have a privileged and intimate perspective on the material and could not be better placed to unfold the most important features of the SSQ experience.

This book will provide readers with a unique angle of reflection on the momentous questions that arise at the interface of science and the moral and spiritual belief systems that we inhabit. In these pages the framework for understanding that relationship is personal stories and "practices," and this seems appropriate. For whether they are technical and theoretical, or whether they concern matters of justice in our everyday lives, our actual practices are a clear measure of what we find most meaningful. They are a mirror—and in the end perhaps the best mirror—of our ultimate concerns.

Practicing Science, Living Faith

Introduction

Jim Schaal and Philip Clayton

This collection of interviews offers glimpses into the scientific and spiritual lives of twelve leading scientists. Here one meets, among others, a British primatologist renowned for pioneering the field study of chimpanzees and respected for advocating environmental protection; a Moroccan astrophysicist who is tracing the formation of stars and galaxies while promoting dialogue between Islam and the West; an African American computer scientist who led in developing the first personal digital assistant, launched a successful software company, and established a foundation for African villagers and American inner-city kids; and an Iranian Canadian child psychologist who is developing novel therapies for neurologically disabled children.

In these pages we will also encounter people with strong spiritual commitments. The molecular biologist—a cancer researcher who teaches biology, religion, and psychology—is an observant Jew; the cognitive scientist, who is American Scottish and works in community mediation and conflict resolution, became a Quaker in high school; and during graduate school the American neuroscientist, who co-created remarkably effective teaching software for language-impaired children, added Christian practice to her Jewish upbringing. The nutritionist, an Indonesian child development expert, is a devout Muslim; the British biochemist, who is producing new drugs to combat devastating infectious and autoimmune diseases, is a lay member of an Anglican religious order, and the Iranian Canadian child psychologist belongs to the Baha'i Faith. The entomologist, who trained a generation of scientists to bring sustainable agriculture to sub-Saharan Africa, was influenced by both African animist and Christian traditions; the cell biologist is an influential writer on naturalistic approaches to spirituality; and the software entrepreneur is a lifelong Catholic and a gospel musician. The astrophysicist is passionately devoted to Islam and also moved by the story of Jesus. The mathematician, a Dutchman who is building the foundations for new computing languages while meditating for world peace, practices vipassana meditation in the Theravada Buddhist tradition, and the Anglican-born primatologist has captivated millions with her stories about signs of spirituality in our primate cousins.

This rich collection—of scientific disciplines, religious traditions, and cultural backgrounds—conveys just some of the ways that scientists practice their science and live their faith.

Reconceiving the Science-Religion Debate

Since the rise of modern science in the West, its relations with religion have been complex. Some have hoped for science to vanquish religion, freeing humankind at last from superstition in the triumph of reason, while others have wished for religion to overcome science, preserving mystery in the revelation of faith. Between these extremes lies a vast number of more irenic alternatives. Matters were never as simple as Galileo versus the church or Darwin versus the creationists.

In his Gifford Lectures the physicist and theologian Ian Barbour proposes a fourfold typology of relations between science and religion.[1] The first alternative, writes Barbour, is conflict, the view that science and religion make diametrically opposed claims that cannot be reconciled but can be settled only by dispute or domination. A second choice is independence, in which the two are understood as addressing different matters or speaking completely different languages. The claims of science and religion may be significant within their own domains, but they have nothing to say to each other. A third possibility is dialogue: science and religion address at least some of the same concerns and hence can be brought into productive conversation. The fourth alternative is integration, where the attempt is made to fuse science and religion, or at least their respective claims, into a coherent whole.

The scientists interviewed here are no strangers to the conflict picture. The biologists have had their exchanges with young-earth creationists, the medical scientists with religiously motivated opponents of genetic engineering and contraception, the technologists with traditionalists and believers in magic. Yet none sees science and religion as bitter enemies locked in mortal warfare. Instead, they experience the conflict intermittently, as skirmishes along the borders.

Some incline toward the independence picture, regarding their scientific and their religious understandings as wholly distinct and unrelated. For these scientists the distinction is not a matter of simple compartmentalization along the lines of "I'm a believer on Sundays and a scientist from Monday through Saturday"; independence is a more carefully delineated boundary between what one can say as scientist and as believer. Most advocate dialogue. They see the potential for fruitful interaction between science and religion but wish also to draw attention to the fundamental differences between the two spheres. Several scientists also seek integration, though none claims to have achieved a complete synthesis of science and religion. Rather, the integration they describe here is of a more limited and qualified kind. Clearly, one scientist's form of integration may be different from another's.[2]

As this book appears, our society stands in the midst of a heated debate about the relationship between science and religion; all signs are that the battles will continue for some time. Much of this debate is concerned primarily with the conceptual relations between scientific and religious *theories*—with the interactions between ideas and truth claims drawn from these two spheres. One finds scholarly analyses, for example, of the relations between physical cosmology and the Christian doctrine of creation or between neuroscience and the Buddhist phenomenology of consciousness.

By contrast, the interviews collected here focus on the *practices* of science and spirituality. Here we observe individual scientists who are seeking integration in their daily lives, even while they recognize tensions between the professional cultures of science and religion. The complexities revealed here resist the sorts of sweeping generalizations that one often finds in the "professional" science-religion debates. Among the SSQ scientists, for example, are atheists who resist reductive materialism, Buddhists who argue for scientific realism, and conservative Christians who favor evolution over creationism. The realities conveyed by these life stories offer an important corrective to the simplistic generalities and formulaic answers one often finds in discussions of this subject.

Practicing Science

This collection does not claim to offer a final synthesis of science and religion. But it does show how it is possible for serious people to pursue both scientific and spiritual understandings of the world and to do so with integrity as well as intellectual rigor. This is not a trivial conclusion, for many still dispute it. Many hold that scientists are supposed to only hold beliefs on objective matters that can be empirically tested, dismissing all else as superstition. Religious beliefs, which seem to be beyond the reach of empirical testing, are said to be subjective and irrational; hence scientists should avoid them at all costs.

Yet science is, after all, a human enterprise. The standards of objectivity and rationality put forth by the founders of modern science in the Enlightenment were, at best, idealizations that masked the subtlety of the realities they sought to demystify; at worst, they were constructs that served to privilege science and its practitioners as the final arbiters of knowledge. In our day influential philosophical and sociological critiques are challenging claims to the fundamental objectivity of science. Examples drawn from the history of modern science itself also raise questions about how thoroughly objective and rational this history has really been—even as they demonstrate the impressive power of the empirical method. On the other side, theological and psychological studies show what an integral role reason and experience play in most religious belief. All these conclusions lead one to suspect that the Enlightenment dichotomy between scientific and religious ways of thinking was far too sharply drawn.

Seen against that backdrop, this collection of interviews casts serious doubt on the oft-repeated claim that scientists cannot be spiritual and that spiritual people cannot be scientists. The interviews present scientists of substance and stature—faculty at major universities, directors of large research institutes, leaders of professional societies, winners of scientific prizes and academic honors, authors of important textbooks and widely cited papers in peer-reviewed journals. Some are theoreticians, others experimentalists and observational scientists, yet others applied scientists and technologists. Several take multiple approaches, moving from chalkboard to observatory, computer workstation to language clinic, field station to laboratory. Some are primarily researchers, others teachers, yet others business entrepreneurs and authors and policy advisers. Whatever the approach or context, they are important contributors to the growth and application of science today.

The range of the lives and stories helps to convey what a diverse set of activities falls under the heading of science. As one sees here, the scientific enterprise encompasses at least five different dimensions. First and foremost, scientists pursue their own research. Research is their defining preoccupation; when their occupation becomes university administration or business management, they typically long to return to their research. What counts as research varies greatly across the fields of science, but in each case the scientist deploys theories, tools, and techniques honed through years of rigorous education, long apprenticeship, and painstaking refinement.

Next, scientists conduct their research within a professional community. They collaborate and compete, create and debate with colleagues in their field. Consider for a moment the field of particle physics. A theorist may interact with a relatively small number of peers in her work, whereas an experimentalist may join a team of hundreds for just one experiment. Yet theorists rely on experimentalists and experimentalists on theorists, and together they depend on a whole host of graduate students, engineers, technicians, and administrators. Since the beginning of modern science, the professional communities of science have been international, drawn together across political borders by societies, conferences, journals, and correspondence. With the burgeoning of new disciplines the journals proliferate; and with the advent of the Internet one can barely keep up with the accelerating pace of communication among colleagues.

Third, scientific communities operate within a complex network of political, social, and economic institutions that surround science. Consider the field of genetics. In the course of a routine week as head of a research group, a senior geneticist may interact with personnel from government funding and regulatory agencies, corporate sponsors, foundation grant makers, and staffs of nonprofit organizations as well as with representatives of academic institutions, professional societies, and publishers. With few exceptions, these institutions grow more market driven (and, paradoxically, more bureaucratic) with each passing year. Fund-raiser, politician, entrepreneur, publicist—one must be all these to succeed in the institutional environment of science.

Fourth, scientists are steeped in the culture of science. Together with others in the same discipline, they celebrate the rites of professional initiation, enforce the ethical and academic standards of their field, admire the great scientific figures of the past, and compete for influence in the present and honors in the future. Scientists tell inside jokes, some self-deprecating and others at the expense of neighboring disciplines. Specialists are bound together by the often highly technical vocabulary of their particular field. Despite important differences and some underlying prejudices, science represents a single culture insofar as all the various subfields are bound together by a shared set of values. These values include intellectual freedom, the public disclosure of methods and results, citation of sources, and, above all, the ideals of testability, or "replicability," and accountability to the data. This distinctive culture of science, famously contrasted by C. P. Snow to the culture of the humanities, may or may not constitute a comprehensive and self-sufficient worldview, with its own sense of significance, standards, and values. Indeed, this is one key question explored in the pages that follow.

Fifth, scientists are inevitably immersed in larger cultures that they experience, from time to time and place to place, as alternately reverent, hostile, or indifferent toward science and scientists. In the present moment of world history some applications of science rouse widespread concern (nuclear energy, weapons of mass destruction, the genetic modification of foods), while others garner widespread applause (vaccines, cancer treatments, AIDS drugs). Still other scientific applications—one thinks of stem cell research, artificial intelligence, and nanotechnology, to name only a few—awaken uneasy suspicions even as they promise progress; we do not yet know what to make of them or how to weigh their risks against their benefits. Nor is the cultural impact of science limited to its technological applications. Consider the historical weight of theoretical contributions by scientists such as Copernicus and Galileo, Newton and Laplace, Darwin, Einstein, Bohr, Heisenberg, Turing, Watson and Crick. Part of the undeniable power of science is to yield results, to alter the prospects for humanity, and to change the face of the earth. This power to yield knowledge lends scientists a measure of authority, both intellectual and moral, in the larger culture. When scientists speak to public issues, people listen.

When one considers the great diversity among these five dimensions of science, one understands why it is important to think of science not just as theory but also as a practice—or, rather, as a range of practices. Too often people conceive scientists as working in isolation: Descartes dreaming up analytic geometry from his sickbed, Newton conceiving the theory of gravitation under the apple tree, Kekule envisioning the benzene molecule in the night, Einstein working out special relativity in a Swiss patent office. In fact, science today can no more be practiced apart from professional communities and institutions than law can be practiced without legal codes and court systems or medicine can be practiced without hospitals and pharmaceutical companies.

Living Faith

If science can be called practice, so too may religion. Admittedly, religious practice has not been the main focus of Western intellectual history. Treatments of Western Christianity, for example, tend to emphasize doctrine. As a result, many think of religion as primarily about beliefs: one is a Christian if one professes the ancient creeds of the church or the Protestant confessions or a sufficiently high belief in the authority of scripture. (The histories of Judaism and Islam show some similar tendencies though perhaps in a less pronounced fashion.) Yet surely religion is not just about beliefs. What, for example, should one make of religion in tribal cultures across the globe, where belief and ritual are often so intertwined as to be inseparable? What of the various other ways that traditions gauge the fervency of individual religious commitment: rituals and sacraments, meditation and devotion, the quest for a lifestyle of compassion? And what of the contemporary trend toward "personal spiritualities" that eschew doctrinal formulations altogether? Religious identity is a complex notion that spans multiple levels of organization, from individuals to broader institutions to whole societies. If one is to do justice to these varied examples, one must also understand religion as practice: a multilayered phenomenon that includes rituals, institutions, and cultures as well as beliefs.

Hence the title: *Practicing Science, Living Faith.* Some scientists interviewed here would be entirely at ease being called people of faith, for they see themselves as standing in an established religious tradition by virtue of culture, conversion, or spiritual discipline. Others, however, might be less comfortable with the label—either because they do not accept the doctrinal tenets of any particular faith or because they understand their spirituality as a matter of experience rather than confession or revelation. We thus intend the term *faith* in the broadest sense possible, as the lived experience of spirituality. One frequently hears the phrase, "I'm spiritual but not religious." Usually, the expression means, "I affirm a spiritual dimension of my existence, whether or not I am actually affiliated with a specific religious tradition." Clearly, it is possible for people to have spiritual experiences, to have faith in something, and to act on that faith—without necessarily believing in or belonging to a religion. Of course, conversely, established religions can also be living faiths.

In the end, then, what these interviews reveal is that the life of faith can be a natural counterpart to the practice of science. One catches glimpses of how leading scientists lead their daily lives, in a way that articles in scientific journals could never reveal. These men and women pray, meditate, contemplate; they sing, chant, or keep silence; they contribute to their faith communities and engage their surrounding communities. A new facet is shown when one approaches the science-and-religion debate from the standpoint of faith lived, not merely discussed.

Where Science and Spirituality Overlap

The scientific and spiritual dimensions of humanity refuse to stay in watertight boxes. Mathematical symmetries in astrophysics, the diversity of life forms in biology, the complex interconnections of brain regions in the neurosciences—each of these areas of study gives rise to a sense of awe and wonder on the part of the scientists who study them. Yet note that awe and wonder are responses more typically associated with the religious or spiritual life. Likewise, humility is a virtue endorsed in many of the world's religious traditions. But it is also a natural response for those who stand before a world of such great beauty, power, and complexity. The practice of science not infrequently gives rise to a natural humility, a humility grounded in knowing that one is preceded by brilliant theorists and experimentalists and that one will be followed by critics astute enough to uncover any weaknesses in one's own work.

Still, beyond awe, wonder, and humility, it is perhaps in the area of *values* that scientific practice and religious practice evidence their greatest parallels. (Indeed, the book might well have been entitled *Values in Science*.) The practice of science, like the practices of law and medicine, is guided by implicit and explicit codes of ethics. The most fundamental of these govern the methods that make up science and the means for reporting it. They demand ruthless honesty in the way that one does research and the way one presents it. The scientist who falsifies hypotheses with sound data may stir a revolution, but the scientist who fabricates data is forever discredited. The strongest safeguard against such unethical behaviors is the requirement that all results be independently replicated before they are generally accepted. Despite the occasional headlines about hoaxes and scandals, the community of science has remained remarkably successful in enforcing its most basic codes of intellectual honesty.

Such intellectual honesty is basic to the practice of science—indeed, it is one of the hallmarks distinguishing science from pseudoscience. But, as these pages show, many scientists are equally concerned with the *applications* of their work, knowing only too well that it can be used for good or for ill. Long before Oppenheimer and the atom bomb, physicists and chemists worried about military applications of their theories, and long before the present debates about cloning, biologists struggled with the controversies about eugenics and social Darwinism. The practice of science, precisely in virtue of its empirical power and theoretical influence, is morally consequential.

Yet the consequences of basic scientific research are not always morally clear cut. The physicists who developed the atom bomb did not think of themselves as trying to kill innocent civilians in Hiroshima and Nagasaki but rather as trying to bring an end to fascism and world war. Worse, the consequences are not always predictable. The researchers who developed antibiotics were working to save lives, but they could not anticipate that the indiscriminate use of antibiotics in medicine and food production would expose their grandchildren to more virulent, antibiotic-resistant strains of bacteria. Beyond the relatively straightforward ethic of intellectual honesty, these larger value questions in the application of science are of deep concern to many scientists: What should I study and not study? How might my work be used for the good of humanity and the world, and how might it be used to harm humans, animals, or the environment?

Each in his or her own way, the twelve scientists featured here are wrestling with the core questions of science and values.[3] Most struggle with applied ethics in their own research: Given a broad respect for the dignity and welfare of human subjects, what experimental protocols will best ensure that no harm is done to them? If my religion teaches that human life is sacred, should I use my expertise to develop fertility treatments for the affluent childless or contraception for those impoverished by overpopulation? These scientists are often interested in applied ethics as a matter of public policy: Is it acceptable for governments to grant multinational corporations the right to patent genes? Is it ethical for pharmaceutical manufacturers to test products on animals? Some scientists are interested in normative questions of conduct: As a scientist working in the private sector, what are my obligations to the academy and to the public? Given the competitive race for a commercially marketable software application or drug or weapons system, to what extent am I required to disclose my methods and results? Finally, many raise the most general questions about value. What is more important: advancing human knowledge or improving human welfare? Which is the greater good: human flourishing or ecological conservation? Although all these questions are framed as "either/or" dichotomies, in actual applied situations scientists must often contend with multiple alternatives and tough choices between several less-than-optimal options.

The scientific dilemmas raise what are called meta-ethical questions: What do we mean when we speak of individuals and the common good, of duties and freedoms, of risks and benefits? Do ethical precepts have universal meaning, or are they restricted to specific cultural and historical settings—or are they merely empty fictions? If we can speak intelligibly of natural laws, can we also speak of moral laws? If so, is there any systematic connection between natural laws and moral laws?

Such questions arise, tacitly or overtly, in many of the conversations that follow. Some scientists seek to explain human moral capacities—whether for altruism or aggression—by sociobiological accounts that involve selfish genes, kinship selection, and other evolutionary mechanisms. Others consider the anthropic principle in cosmology and wonder whether it might indicate anything morally significant about the place of humankind in the universe. Because science explores realms that are not yet well understood, some of these ethical questions remain necessarily speculative. For example, if we find evidence for sentient extraterrestrials, should we expect them to have moral agency? Still, given the pace of scientific progress, today's thought experiments may tomorrow become debates in applied ethics. If we succeed in building robots with artificial intelligence similar to our own, for instance, should we treat them as if they have the rights and values of people?

The Science and the Spiritual Quest Interviews

The scientists interviewed here participated in Science and the Spiritual Quest (SSQ), the seven-year program of the Center for Theology and the Natural Sciences that was funded in large part by a generous grant from the John Templeton Foundation.[4] By the conclusion of its second phase in 2003, the SSQ program had gathered 123 leading scientists in private workshops to discuss the connections they saw between science, spirituality, and values. SSQ's first groundbreaking public conference, held in 1998 at the University of California–Berkeley, brought the private dialogue into the international media spotlight. Between 2000 and 2003 the SSQ private workshops involved sixty-five scientists from seventeen nations, representing a variety of Hindu, Buddhist, Jain, Baha'i, animist, pagan, and nature-centered practices as well as diverse streams of Judaism, Christianity, and Islam.

The mission of Science and the Spiritual Quest was to foster dialogue between scientists, and with the public, concerning the connections and disconnects, the agreements and tensions, between two of the most powerful forces in human culture today.

The Science and the Spiritual Quest program held sixteen public events in nine countries on four continents, together reaching more than twelve thousand people in on-site audiences and touching millions more through media coverage. From Harvard University to the National Institute of Advanced Studies in India, the scope of SSQ zoomed from the grandest pictures of the cosmos to the most intimate portraits of human nature. With meetings at UNESCO World Headquarters in Paris and Gakushuin University in Tokyo, the SSQ conferences focused on deep metaphysical questions and pressing ethical concerns. So unusual was it for Nobel laureates and other eminent scientists to speak publicly of their religious convictions and questions that *Newsweek*, *U.S. News and World Report*, *Der Speigel*, *Le Monde*, the *Times of India*, along with many other major news outlets, ran cover stories on Science and the Spiritual Quest.[5]

In each case the SSQ process began with interviews. Twelve of the most memorable discussions are collected here. In general, the interviews move from the participant's scientific background and current work through her religious background and current interests toward an exploration of any connections she might see between them. The scientists are variously assertive or reticent, direct or meandering, sober or jesting. The interviewer invites reflections, draws out lines of reasoning, probes doubts, and occasionally brings seemingly tangential comments into focus.

Since they were conducted and transcribed, the interviews have served a series of important functions. In the SSQ workshops they introduced each scientist to his peers. It is difficult to be overly formal and cautious when some of one's most intimate thoughts about science and religion are sitting on the table in front of all the other participants. In actual practice these interviews gave rise to some of the most intense, probing, and open discussions that one can imagine in a roomful of people. For three days small groups of scientists from the same field explored the implications of these interviews. They prompted the scientists to articulate, often for the first time, their views on science, religion, and their core values. Out of these first formulations arose an extensive series of papers on science and spirituality, many of which have since been presented in forums and conferences around the world and some of which are now published.[6]

For readers who did not attend the SSQ workshops, the interviews will demonstrate the distinctive ways that scientists approach the comparison of science and religion or spirituality. Topics and themes that rise to the surface here are different than those that arise in discussions among philosophers or theologians. Of equal importance, the interviews provide glimpses of the scientists' innermost thoughts and feelings and a window on important conversations that only a small number of people were privileged to attend.

In editing the interviews we have strived to preserve their conversational tone and to allow the scientists to speak fully in their own voices. For many of the scientists English is not a native language. Thus one must imagine the clipped accent of a Dutchman, the gentle lilt of a Moroccan educated in France, the refined English of an Iranian psychologist trained in Britain. Although we have edited all the interviews for content and length, we have sought to preserve not only the views but also the style and persona of each scientist. A conversation is a snapshot of a person's views at a particular time, and the speakers would not necessarily put things in exactly the same way today. Still, a good snapshot can also catch something timeless in human experience. We believe that these twelve interviews provide accurate verbal pictures of some of the basic ways that scientists connect their lives in science with their lives of faith.

NOTES

1. Barbour's Gifford Lectures were originally published in two volumes as Ian G. Barbour, *Religion in an Age of Science* (San Francisco: Harper and Row, 1990), and Ian G. Barbour, *Ethics in an Age of Technology* (San Francisco: HarperSanFrancisco, 1993). The first volume was revised and expanded in Ian G. Barbour, *Religion and Science: Historical and Contemporary Issues* (San Francisco: HarperSanFrancisco, 1997). For a popularly accessible account, see Ian G. Barbour, *When Science Meets Religion* (San Francisco: HarperSanFrancisco, 2000).
2. Thus it perhaps is more accurate to speak of a quest for "hypothetical consonance"; see the typology presented by Ted Peters in Peters, ed., *Science and Theology: The New Consonance* (Boulder, Colo.: Westview, 1998).
3. See the classic 1956 treatment by Jacob Bronowski, *Science and Human Values* (New York: Perennial Library, 1990).
4. Established in 1981 by the physicist and theologian Robert J. Russell as an affiliate of the Graduate Theological Union in Berkeley, California, the Center for Theology and the Natural Sciences (CTNS) is a nonprofit membership organization seeking to promote the creative mutual interaction of theology and science. Among the premier research institutions in the field of science-religion dialogue, CTNS stands at the forefront of a movement that has seen similar organizations spring up at universities, colleges, and seminaries around the world. (For more information about CTNS and its current programs, visit its Web site, www.ctns.org.)
5. See, for example, the cover article "Science Finds God," *Newsweek,* July 20, 1998.
6. See W. Mark Richardson and Gordy Slack, eds., *Faith in Science: Scientists Search for Truth* (New York: Routledge, 2001); W. Mark Richardson, Robert J. Russell, Philip Clayton, and Kirk Wegter-McNelly, eds., *Science and the Spiritual Quest: New Essays by Leading Scientists* (New York: Routledge, 2002); Roddam Narasimha, B. V. Sreekantan, Sangeetha Menon, and Philip Clayton, eds., *Science and Beyond: Cosmology, Consciousness and Technology in the Indic Traditions* (Bangalore, India: NIAS, 2004); Jean Staune, ed., *Science et quête de sens* (Paris: Presses de la Renaissance, 2005; English translation: *Science and the Search for Meaning* (Philadelphia: Templeton Foundation Press, 2006).

1. Jane Goodall

Jane Goodall is a world-renowned expert on chimpanzee behavior and advocate for animal rights and the environment. One of the first women to do fieldwork with animals in Africa, she began working with chimpanzees in Tanzania in 1960 with the support of the anthropologist and paleontologist Dr. Louis Leakey. Her research at what would become the Gombe Stream Chimpanzee Reserve helped revolutionize ethological and anthropological methodology. Indeed, Jane Goodall's work has fundamentally changed the way the world looks at animals. Defying scientific convention, Goodall named the chimpanzees she studied. She observed their

---

Philip Clayton interviewed Jane Goodall in New York at the State of the World Forum, and W. Mark Richardson interviewed her in Bangalore, India, at the National Institute of Advanced Studies. Bonnie Howe edited the interview for publication.

behaviors with the eye of an anthropologist in search of personalities and emotions, rather than numbering the animals and distancing herself from the species. Research at the reserve is now supported by the Jane Goodall Institute, which also funds other research and education and conservation programs.

Though Goodall's first love and unique gift is her relational research and interaction with the chimps of Gombe, her concern for their welfare takes her far from that home base. She travels as many as ten months a year, speaking on behalf of the animals and their ecosystems, whose continued existence is threatened by human encroachment and industrial and agricultural development. Goodall continually urges her audiences to recognize their personal responsibility to make change. Indeed, her own life has become an exemplar for what can be accomplished when individuals take on this responsibility. Her honors include the Medal of Tanzania, the National Geographic Society's Hubbard Medal, the prestigious Kyoto Prize, and the third Gandhi-King Peace Award for Nonviolence. She has been named a United Nations Ambassador for Peace and was a member of the advisory panel named by U.N. Secretary-General Kofi Annan to promote the goals of the World Summit on Sustainable Development, held in Johannesburg in 2002. In a famous passage from *Reason for Hope,* she describes where her sense of personal responsibility has led her:

> Each one of us matters, has a role to play, and makes a difference. Each one of us must take responsibility for our own lives, and above all, show respect and love for living things around us, especially each other. . . . I share that message in my lectures around the world, with all kinds of audiences and especially with children. I have many definite goals for the future. An important one is to build up an endowment so that our work in Africa—Gombe, the sanctuaries, and

> our programs to help the villagers—can continue in perpetuity. I want to put much effort into spreading Roots & Shoots around the globe, strengthening it, encouraging, motivating and inspiring our youth. And always I have this feeling—that I am being used as a messenger. (266–67)

One hardly needs to describe Jane Goodall's appearance: pictures of her with the chimps of Gombe are icons throughout the world. Thanks to the many televised documentaries and the IMAX film *Jane Goodall's Wild Chimpanzees* (2002), many readers will be able to hear her voice as they read the words that follow. Goodall speaks in a cultured British accent. She is sharp, articulate, and intellectual; her manner is engaging and accessible. The words flow easily from her, especially when she tells stories about chimps in the wild—something she loves to do. Her personal warmth, felt by audiences around the world, is even more intense in a one-on-one discussion. She speaks with utter honesty and sincerity, even youthfulness, removing all formality or stiffness from the discussion. Above all, in speaking with Jane Goodall one feels her passion for the animals and their environments. It shines in her face when she describes them; it quivers in her voice when she talks of what injures them; and it radiates through her clear eyes in every sentence she speaks.

What is the relationship between your spirituality—the things that motivate you at the deepest level, the things you care most about—and your having become a scientist? As we've talked with dozens of scientists about this topic, we've noticed that they describe three different sorts of motivations. Some scientists start with a sheer love of knowledge: they love to study, and that study turns them toward the sciences. Others were inspired by some kind of vision, something they cared about that led them into science. And still a third group tries to pull science and religion or spirituality together into a unity. Which group do you fit in?

It's very simple, really. I was motivated from a very early age to go to Africa and live with animals and write books about them. It was a mixture of absolute fascination and love of animals and a desire for the unknown, all the wild places. All my childhood books were about Africa and the Amazon jungle and that sort of thing. Eventually, I got to Africa and met Louis Leakey. He gave me this amazing opportunity to go and learn about chimps. When I'd been there for about a year, he wrote me to say, "You have to get a degree, because I won't be around to help you get money forever. You've got to stand on your own two feet. You must have a degree for that. We have no time for a B.A., you'll have to go straight for a Ph.D. You'll be studying ethology at Cambridge." I didn't even know what ethology was! I was so scared! I was totally naive, and I had no way of looking it up in the field. It wasn't until I got to England that I realized what "ethology" was. I was dropped in at the deep end of science, you might say. It came as a huge shock, because I'd done everything wrong.

So your love of animals and your passion for understanding them motivated your getting into the sciences. But then, when you officially entered the academic scientific world, when you entered the degree program at Cambridge, people began to tell you that you had done everything wrong. What were the standards that were held up in the scientific community—the standards by which your work was judged "wrong"? What sort of self-identity did you take on when you were starting out as a scientist? Perhaps you could tell the story of how your understanding of science changed in the process of your own work.

My story makes it clear how much transformation there has been in the science in which I got my Ph.D., ethology, and in the status of women in scientific research. The first standard I violated was the standard of basic science education. When I went out to begin studying chimpanzees in 1960, I had taken no science courses. I had not been to university at all. At my high school I had wanted to study zoology, because I wanted to learn about animals. But it was a small school and only offered biology, so that was what I studied. I had a passion for learning about animals in Africa that led me to read all the books that I could.

When Louis Leakey gave me this opportunity to go and study the chimpanzees, it was very hard for him to get the money, because I had no training. It was also very hard for him to get the permission, because I was a girl. That is the second criterion I failed to fit: the male gender expectation. Girls didn't go tramping into the bush studying animals. Actually, few men did it, either. It was considered too dangerous for a girl, so I had to have a companion. That was the stipulation of the British authorities in what was then Tanganyika. For the first four months my mother, who had supported everything that I had done, came with me. Eventually, the authorities decided that I was crazy but that I wasn't harming anything. I had two Tanzanians in the camp with me. Initially, I was made to go out in the field with someone, but gradually I managed to get out of that too.

So there I was at Gombe. Nobody had been out in the field with chimpanzees before, except for a two-month study in west Africa by Henry Nissen. He tramped around in the field and the forest with eleven porters, so obviously he didn't see many chimps.

The first problem I had was that the chimps ran away. There were no guidelines laid out for this study. I had my own intuition: I decided I would never hide from the chimps because it wouldn't be possible. I had gradually to get them used to me. I invented my own ways of doing this, sitting up on a peak with binoculars and not getting too close too quickly. When there was a fruiting tree, I made a little blind, not to hide from the chimps but so that they would realize I was always there. I didn't come out of it; I stayed in that little blind. Eventually, they came to accept that I wasn't a threat to them.

What you were doing at Gombe was so different from what was done at Cambridge in the early sixties. Can you explain a little bit about how your methods stood out in marked contrast to the standard mode? How did the scientific community at Cambridge receive you?

As I have said, I arrived at Cambridge to do a Ph.D. with no B.A., and there was a lot of resentment: "Who is this young girl doing some crazy thing with animals in the field and talking about the chimpanzees as personalities? Giving them names!" Why hadn't I given them numbers?

What was going on at Cambridge at that time was basically ethological studies of, for example, canaries—learning about their song. To learn about their song you deafened them. Other ethologists were learning about the early behavior of domestic chicks—how they followed their mother and how they learned how to peck for food. They were deafening hundreds of chicks to see what happened if they couldn't hear the clucks of the others. Other scientists were working on behavior in insects and tapping into their brains, attaching electrodes to the hunger area of the brain. The insects would eat until they exploded.

The only study being conducted at Cambridge that was remotely akin to mine was the study of rhesus monkeys. The main study there, directed by Robert Hinde, involved taking babies away from the mothers and keeping them away for a specific amount of time. Robert was interested in separation and in the resulting depression. He related it to children's being left alone in hospitals. He was my supervisor. Once he actually went to the home office that gives permits to do research on animals, and he told them that they *shouldn't* have given him a permit because it was cruel. He still did the research, but at least he got people thinking that way.

What would motivate highly educated and civilized scientists to have this understanding of ethological studies? What was the motivation for these ways of treating animals, which strike us today as cruel?

Before I talk about possible motives, I must tell you that these practices are still basic methods of ethology. They are still deafening songbirds; they are still blinding kittens. Marc Bekoff and I just started an organization called Ethologists for the Ethical Treatment of Animals. It's something I've wanted to do for a very long time, because to this day ethologists do horrible things to animals.

It sounds like the motivation is to get rigorous scientific knowledge of the animals. Is the justification for the cruelty that there is no other way to acquire this knowledge?

Yes. Some of the research is related to human problems. The separation of the monkeys was designed to encourage hospitals to let parents stay with their children. That was very humane—to the children but not the monkeys. Some of it, sadly, is motivated by the desire for knowledge that might or might not be applicable.

Many scientists seem to believe that knowledge is so important and so valuable for its own sake, apart from any particular use it might have, that how we treat the animals along the way simply isn't important.

Yes, there is still a legacy from the animal behaviorists of the late fifties, who maintained that animals were simply bundles of stimulus and response. When I got to Cambridge University in 1961, there was an emphasis on invasive experimental work to find out how animals operate. My ways of describing chimpanzees and their behavior were perceived as shocking. I described their personalities, minds, and emotions. And I gave them *names* rather than numbers.

That fact alone is so significant. To number chimps is to say, "They're all machines; let's see how this one will function." When you chose to name them, it expressed some insight or intuition that was fundamental to your entire work. What was that insight?

Let me run through the whole thing and then come back to your question, because then it will make more sense. I named them; that was the first mistake I made. This naive young girl, what can we expect of her? She doesn't know anything. I had named them. Then I talked about their different personalities. How David was so different from Goliath, and Goliath from William, and Flo from Olly. That was not permissible. Probably, there were individual differences in nonhuman animals, but they were best swept under the carpet. That's what I was taught: I wasn't to talk about personality. That was a uniquely human characteristic, and I was told in no uncertain terms not to talk about it.

The entire assumption of animal studies was that because they are not human, they do not have the sorts of differences that we have?

That's right. One couldn't talk about personality. One shouldn't even talk about *individuality* in animals. The goal was to describe the characteristic behavior of the *species*.

So the unit of analysis was the species, not the individual. Did your defying that parameter affect your ethological research?

Absolutely. Not only did I focus on individuals and on animal *social*, relational behavior, I dared to suggest that chimps had minds that could make decisions. For example, a chimpanzee is sitting and looking around. He scratches his head, then moves off purposefully. He picks a twig, strips off its leaves, and, holding it in his mouth, goes off to a termite heap that is out of sight. There he uses his prepared tool to fish for termites. Clearly, he was planning to search for termites when he selected his tool.

At Cambridge I was talking about the chimps' having minds. And the worst thing I did was talk about their having emotions similar to those we call happiness, sadness, and so on. Even if I chose to give them names—even if that was eventually accepted by some established ethologists—I still shouldn't talk about emotions. That was the ultimate sin, the worst kind of anthropomorphism—attributing human characteristics to animals.

The dominant paradigm assumed that emotions were only for beings like us, people with minds and conscious will. Since research animals routinely were numbered, treated like machines, they could not have minds. Did the established scientists think you were cheapening humanity or that you were ascribing too much to animals?

Both. I was not conforming to professional, ethological protocols. And I think it seemed that I was lowering humans or else I was elevating animals, one or the other. It didn't matter. It was wrong; both moves were bad. I couldn't do either of those things.

Just for the record, this is not just Cambridge? This was the pattern in the study of ethology at the time?

Yes, in ethology and psychology, that was the pattern. There was this very sharp line that Western science, and to some extent Western religion, had drawn between "us" and "them," humans and animals. The fortunate thing is that Louis Leakey deliberately picked someone whose mind was, as he put it, "uncluttered by the reductionist attitude of ethology." He didn't tell me that, but that was his intention in bringing me in.

To knock over the tables.

Right. But he didn't warn me about this.

At the time it must have just been mysterious to you why these established, seasoned scientists were opposed to things that seemed obviously crucial to understanding animal behavior.

The fortunate thing was that all through my childhood I had watched animals, from one-and-a-half years of age. My first real animal experiment—and looking back on it I find it amazing—was when I was four-and-a-half years old. I loved animals but was living in a city, in London. I went to stay for a month on a farm, and it was magic. Farms were proper farms in those days. So here's this little girl, loving animals, having close-up contact with cows and pigs. It was wonderful. One of my jobs was to collect the hen eggs. Sometimes they laid them in the hedges. But they often laid them in small wooden henhouses where they slept, with a little wooden gangplank that was pulled up at night to keep the foxes out. There were little nesting boxes at the side. I was collecting the eggs and putting them in my little basket, and—there's the egg [*she makes an egg-sized circle with her finger and thumb*], but where is the hole on the side of the hen for the egg to come out?

I was looking at the size of the egg and looking at the hen, trying to work out how this egg could come out of the hen. I was asking everyone, and they apparently didn't explain it properly. This led to my first really, really vivid memory ever. A hen climbed up the gangplank into the little henhouse, and I climbed after her. Well, of course, you don't want somebody crawling after you when you're going to lay an egg. "Squawk!" she flew out. So—and this is what is so amazing—I must have realized that the henhouse would now be perceived as dangerous by the other hens because that hen had left, squawking in fear. And so I crawled into an empty henhouse. My family had no idea where I was and, as darkness fell, they called the police. They had hunted high and low. It seems I waited in that small henhouse for approximately four hours.

I remember that I didn't want to frighten the hens. I knew that the hens would be going in about that time to lay, so I went to an empty henhouse and waited. That was the amazing thing. I waited about four hours; for a child of four-and-a-half that's pretty amazing. I can still see the egg coming out.

So you were successful in your first scientific quest?

I was successful. She came in. I was crouched at the back, hidden in the straw, and I can still see how she turned away from me, rose a bit on her legs, and then a slightly soft white thing plopped down. The excitement of that!

I had this amazing mother, who, when she saw me finally—this little thing all covered in straw—she didn't reprimand me, she didn't say, "How dare you do that?" She saw my shining eyes and sat down to hear this magic story of how a hen lays an egg. As I got older, she got me books about animals, *Dr. Doolittle* and so on.

But when I got to Cambridge, with no preparation for science at all—except knowing about patience and having intuition about what animals will do—I was told that animals were just machines and they should not have names. They did not have personalities and they did not have minds. I knew this was rubbish. I knew it was rubbish, in part, because of my dog. All through my childhood I had this amazing teacher in animal behavior, the best teacher any child who loves animals could have. I knew that Rusty had a personality and was different from other dogs. I knew he had a mind. I set him problems and he worked them out. I certainly knew he had emotions; it was very obvious when he was happy or sad. Everybody knew that at the time, even the scientists who were talking about animals' being bundles of stimulus and response. Deep down, they knew that there was more going on.

Intuitively, anyone who has a relationship with a pet—not to mention anyone who observes animals in the wild—knows that animals are more than stimulus-response bundles. But ideology creeps in and forces people to turn away from that intuition and from the evidence—all in the name of rigorous science. How do you understand this mistake? It is not just a mistake of the early 1960s; in many ways it's just as widespread today. Help us comprehend what leads ethologists to make this error.

The problem is not confined to the discipline of ethology. In both ethology and psychology the same kinds of things are done for the same kinds of reasons. The electric shocks, the deprivation of food and water, the electrodes in the brain—

It sounds as if you are saying that the researcher has to be open to having a relationship with the animals. If the scientist is open to the possibility of relationship, then the animals that she is in contact with will be themselves, and the requisite mutual learning can occur. It's almost as if many people who go to study animals close themselves to that relationship, as if I were to close myself to you and treat you as just a category or a stereotype. Is that kind of categorization and a priori closure to relationship part of the research design problem you are addressing?

I think it's probably the other way around. People get caught up in their desire for knowledge. I hate to say this, but they also get caught up in desire for a certain kind of academic career: credentials, Ph.D., papers. Science is high status, and therefore it is a good thing to do, and you have to carve out your own niche. So, you take something that nobody knows anything about and find out about it. At Cambridge in the sixties, to have any kind of acceptable science you had to be extremely rigorous, so you had to really go against what is logical, what makes intuitive sense.

If a researcher ascribes to the dominant paradigm, and she is trying to learn about a big-brained animal, and she intends to use information from that study to help her understand the *human* creature, it does not make sense to use the animal's body alone—its blood, its immune system, its central nervous system. Experimenting with the physiology is perfectly scientific. But it is inconsistent to experiment in that fashion only. To select certain animals for study because they have "big brains," while denying that the animal subjects' brains are also capable of mental performances similar to ours, is illogical. Some scientists perform horrible psychological and ethological experiments on animals, trying to learn something about the workings of the human brain, and yet maintain that these same animal subjects have no feelings or emotions. That can't be. The commonsense observation that all big-brained animals share with us certain capacities, among them, the ability to feel and to make decisions—that's what these methods deny.

It's really active denial, isn't it? Because the much more instinctual response that we have as animals is to build connections with the other animals we encounter. Every child knows that. But there's a process of unlearning that science seems to have fostered.

Yes. The refusal to admit that other animals share with us feelings and sentience has led to incredible cruelty. Taken to its extremes, it has led to the dehumanization of certain groups of humans, as in the Holocaust. The Jews, the Gypsies, the mentally ill were considered nonhumans. This led to unspeakable atrocities. And this is what we are doing to animals: committing atrocities. Clearly, we have no right to cut up sentient beings. But if you just believe that they are *not* sentient, believe they have no emotions, no feelings—then it is okay to conduct invasive experiments, imprison them, and so on.

We justify the dissection or vivisection by constructing categories. We impose the categories onto these animals and insist that there could not be a relationship, that they are just objects.

Yes. The standard line is that it may *look* as though they have feelings like ours, but that's not so and you can't prove it, therefore it's not true. Science won't approve it if you can't prove it.

The standard for rigorous knowledge is partly to blame? You have described a sort of moral failing, but you have also described an aesthetic failing—to hold up an overly rigorous standard of proof. If science is a slave to that standard, then it will be forced to engage in these unethical practices. What would happen if we let go of that rigorous standard of proof?

Change has happened gradually, and it hasn't happened everywhere. Forty years later there are still pockets of strong resistance to changing the dominant paradigm. I'm so amazed at how strong those pockets of resistance are in some places, particularly in parts of Europe such as Germany and France. But by and large there's been a huge change. It's very fashionable to get a Ph.D. or even do postdoctoral research on animal emotions these days. There are more and more scientists who accept that animals are able to reason.

More than anything else, chimp studies have helped to show that this is so. There is more and more information coming in to show that chimps are capable of intellectual performances that we thought only we were capable of. They have thinking, rational minds. They have consciousness of self. As, one by one, these barriers that we built up to make us unique are broken down, I'm getting more and more questions from my audiences that I didn't have before—questions about spirituality in our species. Perhaps *this* makes us unique.

Spirituality as the unique human attribute—I'd love to hear about that. Perhaps we could start with the answer that you give when that question is raised.

It is my belief—and because it is a belief, you can discuss it but not disprove it—that there is a great Spiritual Power and that there is a spark of that spiritual power within each of us. And I believe that there is a spark of the same spiritual power in *all* life. The one thing that does distinguish us from all the rest of the animals is our spoken language. We have rationalized our feelings about the great spiritual power—God, the Divine, the Creator—whatever we want to call it. We have different names and theories, but it doesn't matter to me what name it has. Some of us call this spark a soul. And most people, even those who love animals, still do not want to think that animals have a soul. We think the soul resides in us alone. Many people are prepared to admit that there is this spark of spirit in everything, but only in us is it the soul.

This human "soul" is a qualitatively different form of spark?

It seems to be. I haven't really talked to many people about it, but this is the sense that I get from people.

Aristotle said what sets humans apart is that we are the rational animals. The soul pertains to the rational animal alone.

But this isn't true.

Right. Because once we see rational, mental, and emotional behavior in the chimps, it undercuts the old dichotomies.

That's right. But there are certain people for whom the possession of a soul is the guarantee of an afterlife, an eternal life. I don't think they are happy thinking of chimpanzees and dogs [as] having an eternal life. But that is Western thinking. Talk to people in the Eastern religions, and they will tell you it's all mixed up; you can be reincarnated as a dog or a chimpanzee or an ant.

Your own understanding of this spark that is shared in all living beings sounds much closer to some of the Eastern traditions, where no firm line is drawn between the human spark, the animal spark, and the sparks of all of living things.

All living things. In fact, some traditions put the spark into water, but I haven't really thought about that. It doesn't seem to matter.

How do we come to understand the connection, the communication from one spark to another? Does this sense that we share something common inside bind us together at a deeper level? What is the capacity by which we might be able to know that? It is not a scientific hypothesis, it is something very different. And yet it is basic to what you do.

Yes, during a dialogue with Jonathan Granoff [a lawyer, author, and peace activist], we talked a little bit about the soul, and Jonathan said, "How can you tell if you look at me that I have soul? Must you cut me open? Will you find a soul?" So I said, "Given that you have a soul and I have soul, I think that I can find your soul by looking into your eyes." When I look into the eyes of a chimpanzee or a dog, I think I can see their souls too. There is something *extraordinary* about eye contact. Once or twice, when I was out in the forest, a snake actually glided over my foot. I knew the snake would not harm me unless it was suddenly frightened. I tried to keep motionless. But once I must have moved a little because suddenly the snake stopped and looked up. And this little reptile, this creature that is *so* different from us, *looked into my eyes*. I've noticed that all kinds of animals look into my eyes—lizards, tiny rodents, birds. I find this utterly fascinating, this establishing eye contact with creatures so dissimilar to human creatures.

It is relationship—not some sort of abstract knowledge but a deeper knowing brought about through this intuitive connection with the animals.

Yes, it seems to be some deep, intuitive knowledge that makes meaningful connection possible via visual contact. After all, these small creatures might just as well look at our fingers or our feet. I suppose they look at these as well. But it's the eye contact that is so amazing.

We have talked about the standard model of ethology, and we have talked about your having broken with that model from the very beginning of your work. You would have lost ten years if you had done an undergraduate degree first.

Yes! Who knows?

What I hear as I listen to you is a beautiful integration of science and spirit. You talk about careful watching, observation and reflection, and at the same time you describe this intuitive sense of connections. It's clear in your decision to name the chimps in your very first fieldwork, and in the sense of intuitive connection that you freely speak of as a spiritual connection with these other living things. If, in a project like Science and the Spiritual Quest, many are trying to overcome a division between the two fields—science and spirit—it seems to me that in your own work the two have always been fully and naturally integrated.

Yes, science and spirit are completely integrated for me. That's why, when you asked the question right at the beginning about overcoming tensions between the two, I said it's a bit different for me. Because there was nothing to overcome, and because I didn't care about the scientific rat race. You can say I was lucky in that regard.

Nevertheless, going to Cambridge was one of the most useful things I did, because I did learn about scientific method. I learned about self-discipline and scientific modes of expression, which I actually enjoy and love. I was absolutely fascinated by the scientific way of expressing myself. I was fortunate to have a mentor with as good a mind as Robert Hinde, who took a very "hard science" approach. I mean, you probably can't understand what he writes about, it's so technically scientific. And yet he was able to step out of that cast to help me express what I was trying to say. I was furious at his criticisms sometimes. I threw my manuscripts in the corner again and again, but then I'd picked them up in the morning and think, "Well, Robert Hinde has a point there." When I wrote that Fifi was jealous because other chimps were touching her baby brother, he said, "You can't say that." And I said, "Well, she was." And he said, "Yes, but you can't prove it." So I said, "Well, what do I say?" And he said, "I suggest you say Fifi behaved in such a way that had she been a human child, we would say she was jealous." Is that not brilliant? So, that was it. That set me on the path I've followed ever since—being careful with how you express yourself so that you don't lay yourself open to being torn apart, because that's not helpful to anybody. I did learn how to present the shocking things I was talking about in such a way that I could be listened to. I couldn't have done what I've done without that discipline. And together we wrote almost a new kind of science, a new twist to the study of animal behavior.

So it was valuable to you to have learned and used the scientific method?

Absolutely.

Because, after all, you could have written novels instead?

[*With a twinkle in her eye*] Yes, I could. I could have gone on being a *National Geographic* cover girl. [*Laughs*]

You objected to the kind of objectification in scientific studies that was justifying cruelty, even brutality, toward animals. Your fieldwork gave you the backbone to advocate a change in the scientific attitude and the methods of your discipline. One could even say you had a "calling" to be an advocate for the animals. What have you gained by seeking to employ the scientific method without falsifying or denying the relationship with the animals? What has this wedding of scientific rigor to intuitive knowledge and a deep connection with your subjects contributed for you?

The questions my fieldwork raised contributed massively to the subsequent changes in scientific thinking in ethology. Nonetheless, I needed to learn how to do scientific thinking and writing. If I hadn't paid my dues and learned the scientific method, I could never have gone into the scientific field with the goal of presenting knowledge in a different way. I wasn't going to stop talking about animals' emotions and their minds. But I completely enjoyed the self-discipline of finding out how to analyze the data so that you could persuade others. When you know something, and then on top of that you can prove it with numbers—that is powerful. These methods certainly had an impact by refining my thought process. Scientific writing, as all writing, takes a lot of self-discipline. Scientific discipline taught me to ask more rigorous questions. I asked different kinds of questions. When I joined the scientific community, I had a whole new group of people to persuade. It was very clear from the beginning that these were people who were guilty of a lot of cruelty to animals, and I knew I had to persuade them that animals are how I know they are. I was right, of course. But I had no right to stand up against those erudite scientists when I knew nothing. If I had gone to Cambridge first, to get a degree, then I'd have gone along with it all, probably.

Did you have a conscious desire to transform the scientific field and the way that it dealt with animals, or was that just a by-product?

Looking back on it, I can see how, right from the beginning, so much of my work was geared to trying to make people see that they shouldn't be doing what they were doing. As soon as I understood a little bit more [of] what the scientists were doing, and how that meant animals were being treated, from then on I had a hope that the chimp studies could help animals in some way. When I wrote that big book on the chimpanzees in Gombe, I had to go back and learn for myself all the things that I never learned because I didn't do an undergraduate degree. It wasn't until its publication, in 1986, that I actually felt like a "proper" scientist. Finally I knew what the people were talking about when they talked about hormones and aggression, brain development, and such factors. I finally learned it—not in huge detail but enough to understand.

Would you recommend the same pattern for a young researcher—to start with the fieldwork and then to go back and pick up the science, the additional knowledge that she needs?

Yes, it's very helpful. Even if it's only a short study of a group of animals around one's own home—either on one's own or as part of a structured summer field course. It is so important to develop an understanding of and empathy with the animals one plans to study. After that, young researchers can develop a much better study plan for their actual research projects. When some field experience comes first, when there is a chance to employ intuitive knowledge, the student will understand something of the animal's essence. Such field experience and chances to deploy intuitive knowledge will shape her future research.

Your approach calls scientists into question—not only from a moral standpoint but also because it asks them to give up some things in which they have some self-interest. It's almost as if you're saying that to give up the traditional approach to science is to give up a self-invested prejudice.

For many scientists, to change their practices would be to open themselves to duty, to questioning their own values and their own ideas. You can see why people want to keep doing research the old way. It is the scientists who do invasive things with animals—and the nonscientists who work with infected farm animals—who offer the strongest resistance to the fact that animals have feelings and personalities.

For these people, to change methodological approaches would probably mean losing their careers. They don't want to do that; they're afraid of it.

Very early on, I learned how hypocritical scientists can be. That was kind of a shock, you know. Once, I recall seeing a scientist whom I had hugely respected stand up to the lectern and say that he believes that aggression is all learned and not innate. Then, when I sat down with him over coffee and asked, "Robert, what do you really think?" he said, "I would rather not talk about what I really think." It was like an idol falling.

By that time I knew that aggression in humans was called violence. I had my baby by then, and I had experienced that surge of absolute, irrational rage that you feel when somebody threatens to harm your child. You could kill someone!

Thus you knew where the rage comes from, that it is biologically motivated.

Exactly. It is as genetic as it gets. Female chimps taught me a lot about the right way to feel with a child. But it was the other way around too. From my experience as a mother I came better to understand the chimps. I suddenly understood why the female would get furious—"Uh, uh, uh!" [*she very loudly, and very realistically, impersonates a female chimp warning away an intruder*]—if somebody just came up, curious to look at the baby. At first, I couldn't understand it, but after I had my baby I could.

Could you teach the chimps, as well as being taught by them?

I didn't try.

It wasn't part of anything you could do?

One can. When one shows them films of mother chimps looking after their babies, they learn from the films. They also learned by having one of the keepers come in and nurse her own baby.

And then they pass those skills on once they've learned them?

Yes. You get these naive males who never learned how to copulate till they watched others ... like pornography!

Can a process of teaching make animals more sensitive?

Yes, it can.

Could you describe one of the more significant spiritual encounters with an animal or group of animals that you've had? Can you think of one of those moments where there was a connection between you and an animal that was special and important to you?

The incident that stands out, and always will, involved David Greybeard. It occurred soon after he allowed me to follow him in the forest, without showing signs of unease. I always tried to follow quietly and allowed him to get away if he seemed irritated by me. On this particular occasion I was crawling after him through a horrible tangle of thorny vines. I remember thinking, "Now I've lost him," because he was able to move through the vines so easily. But I found him sitting, almost as though he'd been waiting. Maybe he was. He knew I was following him, so if he'd wanted to avoid me, he easily could have. Whether he actually, deliberately, was waiting, who knows? Maybe it was interesting for him. Maybe I was interesting. But he certainly knew I was totally harmless. Anyway, he was sitting on the ground and there was a bright red palm nut, which chimps love, lying on the ground. I picked it up and held it out to him. He didn't want it, so he turned his head away; but, pushy me, I moved my hand closer. [*To the interviewer:*] Hold your hand out, and I'll show you what he did. You've got the nut on your hand, and you push it closer to him—like this [*she demonstrates*]. He then looked directly into my eyes. He took the nut out of my hand and dropped it, just with one movement—like that. He didn't want it, but then ... [*she very gently squeezes the interviewer's hand with her fingers and releases it*]. That's chimp reassurance.

That is beautiful.

David and I communicated as our ancient primate ancestors might have. It reached back to our human primate heritage of communicating pre*words*. Using a "language" that preceded spoken language.

As if he were saying, "I don't want it, but it's nothing against you, I'm not rejecting you."

That's right.

That is a beautiful story.

That moment still stands out after all these years. I sometimes talk to kindergarten kids. It's not my thing to do, as they never seem to be listening. They are listening—I know that for a fact—but they're always wriggling about. Anyway, I was telling them this story, and I asked one child to hold out his hand to demonstrate. We had a little nutlike thing, so I could demonstrate. And after I demonstrated David Greybeard's action, I saw all their hands reaching out to me. They all wanted to know what it feels like for a chimp to hold your hand. They'll never forget, and it will probably blur in their minds, because I said, "Do you want to know what it feels like when a chimpanzee holds your hand?" They'll remember the feel of a *chimpanzee* hand!

And they all reached out their hands?

I suddenly realized I had to do it to each one of them.

That's beautiful. What is so powerful about that story is that we do exactly the same thing. A human mother will reassure her child, or a friend will reassure a friend, with the squeeze of a hand; it does as much for us, as primates, as it does for them.

Chimpanzees kiss, embrace, hold hands, pat one another on the back, and tickle; they swagger upright, they punch, they kick, they hit, they pinch. And they do all these things in the same kind of context as we do. They mean the same kinds of things. So even if you are totally naive, if you watch chimpanzees, particularly young ones, interacting, you can pretty well understand what they're doing and why they're doing it—what each of the different gestures means. This was another thing they didn't like at Cambridge: I talked about the chimps' kissing.

They didn't like to hear this part of your fieldwork because kissing is too human? Perhaps because it's too intimate; it involves emotions and feelings and care and concern—and we can't attribute those sorts of feelings to animals?

Exactly. I couldn't talk about embracing—but I did, you see. The first scientific paper I wrote, after I'd been in the field for a year, I submitted to *Nature*. In it I had described some actual observations, like one of

David Greybeard using grass tools, and another where David Greybeard reached out to pat and reassure Flo. Well, *Nature* didn't object to the names, but the editor sent back comments, and every time I had put *he* or *she,* that was crossed out and replaced with *it.* And everywhere I put *who,* that was crossed out and replaced with *which.* So I thought, now they're trying to deny these beings their gender, which is so very obvious. Yet they left the names.

So "David" has to be an it?

David has to be an *it.* So I crossed out the *it*s and *which*es and put *he, she,* and *who*—and that was my first victory. Here's another example. Back then, I couldn't talk about adolescence; that was a uniquely human, culturally defined phase of the life cycle.

I know the arguments in cultural anthropology that support that restriction: adolescence varies so much across human cultures, hence it must be a purely human phenomenon. So they argue, at any rate.

That's the argument, but there are also many physiological changes, which can be measured scientifically, that go on during adolescence. We talk about adolescence in animals all the time now. Childhood was another no-no, and you could not talk about a baby. You couldn't talk about motivation, and you couldn't talk about excitement, social excitement. It was really strange to me; it was a very weird, wrong way of thinking.

You found it inconsistent too? Some data were admitted, while other data were not allowed, which makes one think that ideology and not scientific observation is in control. But I wanted to ask you: How will it change our understandings of ourselves as people, if we acknowledge our relationships, and similarities, with our animal cousins?

I think it gives us a better handle on why we have some of those strange characteristics that we have as humans. We learn a lot when we realize that we are related to animals in this way and that some of the things we do today are not very useful any more, although they were useful in the past and they still are to the chimps.

If we come to see ourselves as related to other primates, without drawing qualitative walls between us, does this awareness have the potential to transform our self-understanding as humans, our sense of ourselves in the world?

Hopefully, it will. I think the biggest impact it can have involves the way that we treat other animals.

That's the transformation that you are working for in the new organization that you and Marc Bekoff have formed, Ethologists for the Ethical Treatment of Animals?

Yes. We seek to promote discussion of the issues. We want to bring some of these issues out and have people from all around the world discuss them. You would be amazed at how much information and how many queries are coming in from every part of the world, from Africa to Russia. It's fabulous. Different ethical questions are being talked about. People are taking courage because they thought they were alone and were being "unscientific" about these issues, but they find that other people feel exactly the same way.

You are hoping to create a network of scientists who share this perspective, a movement of people who already have these convictions but haven't known that others share them? Is the idea that understanding our relatedness to our animal relatives will begin to change our behaviors toward them?

Yes.

Perhaps such an organization of scientists also has the potential to transform what it means to exist as human on this planet. Is there a deeper sort of tranformative potential in your work? In a sense, you've always had this perspective, so perhaps you can't answer the question. But many people are now discovering it for the first time.

I think if it really works the way it should work, this sense of connectedness with other animals should lead to a much greater concern about what we're doing not just to animals, as individual beings, but also to the places where they live. We are so greedily and arrogantly taking and taking, eliminating animals but also the entire way of life of indigenous peoples. In the developed world farmers are losing their land. Did you know that in England there was a ruling by a court that farmers must keep their animals quiet at night? Because people moved out of the city for the quiet of the country!

In our conversation I've gotten the sense of a spiritual vision that underlies your entire work as a scientist—from the story about observing the hen when you were four and a half, right through to your current work with Marc Bekoff in founding this new organization.

Yes. And don't forget the program of the [Jane Goodall] Institute, Roots & Shoots, which is focused on the young generation growing up. We want them to understand why we value animals and the environment the way we do.

The things you do grow out of your scientific study and a lifetime of publications. But they also seem motivated by something that many today would call a spiritual vision, a spiritual understanding of the world, a higher calling. Are you comfortable with those terms?

I'm quite comfortable with them, but I'm not quite sure how to answer your question. Perhaps that's because I think I began with that worldview, and all that I read and experienced from when I was a child reinforced this worldview.

It's hard for a fish to describe water. Could you try to put words on that view of the world—this view that you've always held, that has motivated and been confirmed by your fieldwork, that motivates you to travel and speak more than three hundred days out of the year? Do you ever pause in the midst of the rush and ask yourself, "Fundamentally, how do I view this interconnected world that we are a part of?"

I think it comes down to a very fundamental question: Are we here by random chance or is there a purpose? Is there a meaning to our life on Earth—or to any life on Earth? I can't believe it's chance. If you believe that there is purpose to our life on Earth, then you have to ask, "What is that purpose?" When one considers this whole long process of evolution, how very simple forms have gradually become more complex, how the brain has become more complex, how this has led to consciousness and, in ourselves, to a sophisticated spoken language, so that for the first time we can discuss ideas—all these exciting developments—then it seems that we can tap into a Power up there, a power all around.

Physically, we have these attributes, these brains, this language, and we get glimpses of the amazing things that humans are capable of. There are people who stand out—geniuses, saints, musicians—who show the incredible heights to which people can rise. Yet you look round at the world today and you also see the horrible mess that we have made: the destruction of the environment, the violence, the greed, cruelty, and war. We need a new perspective. We need to think of the millions of years of evolution and what they have made it possible for us to do. First came physical evolution. At some point came the evolution of language. Once we humans had developed a sophisticated spoken language, cultural evolution became ever more important, which means the ability to pass behavioral traditions from one generation to the next through observational learning. With spoken language we can teach our children about things that are not present, discuss the past, make plans for the distant future, discuss things that matter—so that they can grow from the collective wisdom of a group of individuals.

We've moved so far along the path that change now happens *fast*. In just fifty years we've moved into the electronic information age. All these changes raise moral questions. Today we are struggling to understand the differing moral values found in differing cultures. What do we do when a certain behavior is acceptable in one culture but not in another? Is there commonality in moral values? I think there is, at least at some basic level.

Finally, we come to spiritual evolution. There are people who have been living spiritual lives for hundreds of years, but in the West, or parts of the West, we seem to have turned away, very deliberately, from such questions. Many try to deny the presence of a spiritual dimension to their lives. At the same time they are involved in a desperate attempt to find a meaning in their lives. Maybe, just maybe, we can turn things around and begin to save what is left of the natural world. Maybe it is not too late to heal some of the scars we have inflicted, to find ways of living in greater harmony with nature. And perhaps we can start thinking about the spiritual needs of our children. They are so much caught up in the materialistic, selfish world of the West—as children are caught up in poverty and war in so much of the developing world. There are not many pieces of Paradise left.

For a child to develop normally, a relationship with nature is terribly important. Their psyches need it. Some children growing up in the middle of an inner city never see nature—it costs two thousand dollars to take inner-city children in a bus out of Los Angeles into the country, just because of the insurance. Children who *could* afford to get out of the cities are spending more and more time in an unreal world, a world of virtual reality. They're being denied the chance to go grubbing about in the earth. My hope now is that we can get children into a relationship with nature—to use the Internet and these amazing miracles of communication as just a little bit of *extra* but not as the main means of learning.

Such relations with nature don't only help in the development of the psyche; they also keep the spirit alive. Something within dies, surrounded by the walls of concrete.

Yes, absolutely. There's an amazing artist, Ernie Bates, who painted the painting *Love in the Ghetto.* It shows three inner-city teenagers who are sitting around a crack in the cement of a sidewalk. Out of the cement a plant is growing. But in the picture they cannot see the beauty, for their eyes are closed. They are blind to it. They can't see or understand nature. My greatest hope is that people might again open their eyes to the incredible beauty and value of the world around us.

2. Hendrik Pieter Barendregt

Hendrik Pieter Barendregt is a "meta-mathematician," a thinker who reflects critically and constructively on the foundations of mathematics, logic, and computer science. His work has contributed not only to the theory of mathematical proofs and computations but also to the development of new functional programming languages, proof-checking algorithms, and mathematical assistant systems. Raised in an atheist family in the Netherlands, as an adult he undertook Buddhist meditative practice—first in the Zen tradition and then in the Theravada tradition—and is beginning new research on the neurological underpinnings of meditative experience. Though skeptical of notions of faith and transcendence in religion—and even cautious about using

---

Philip Clayton interviewed Hendrik Pieter Barendregt; Jim Schaal edited the interview for publication.

the word *spiritual*—Barendregt sees strong parallels between the mathematical and the meditative "minds."

Born in Amsterdam, Barendregt was educated at Montessori schools, whose pedagogical methods in mathematics encouraged him to pursue the subject at a higher level. He specialized in mathematical logic and completed his doctoral work in 1971 at Utrecht University in the Netherlands. A student of Georg Kreisel's, Barendregt wrote his dissertation on the typed lambda calculus, a mathematical description of proofs and computations that turned out to have far-reaching implications for computer programming.

While doing postdoctoral work at Stanford University, Barendregt came into contact with his first Buddhist teacher—Kobun Chino Roshi—with whom he began training at the Tassajara Zen Mountain Center near Big Sur, California. Upon his return to Utrecht, Barendregt began teaching philosophy of mathematics and publishing in the field of lambda calculus. In 1986 he moved to a new position as a professor of mathematics and computer science at Holland's Nijmegen University, where he became increasingly interested in computer programming languages. Also at Nijmegen he met his present Buddhist teacher—Phra Khru Kraisaravilasa Mettavihari—with whom he has been studying vipassana meditation. Barendregt is writing a phenomenology that connects information obtained through intensive meditation with scientific information obtained through research in neurophysiology.

Conversations with Barendregt are apt to be memorable. His closely shaved head suggests an ascetic bent, while his clothing hints at a comfortably rumpled way of being in the world. More remarkable yet are his eyes: intensely blue, piercingly observant, and crinkled at the corners with years of easy laughter. Speaking in crisp English with a charming Dutch accent, Barendregt conveys some of the same qualities of thought that he has designed into his functional programming

> languages—wide ranging, well ordered, consistent, and precise.
>
> "Do spiritual scientists feel responsible for the results of their work?" he muses. "I think we do, in spite of the fact we often are not! It is commerce, politics, and the military that most often make misuse of scientific results. What we do have responsibility for is to inform the public and politicians about the potential uses and misuses of our science."

Often we find that scientists' intellectual pursuits can be traced back to earlier experiences, even those from childhood. How would you describe the development of your interests?

Well, I attended a Montessori school from the age of four until seventeen. Very important in Montessori schools is the so-called self-control of error. Each child can decide whether what he or she did is correct or not. In this way you do not become dependent on the judgment of an adult, which Montessori claims is denigrating for a child. You also get a reward from the work itself, and that's how you start loving to work. That has been very important for me. I remember very well, for example, some "Aha!" experiences. For example, my father taught me how to add large numbers. I could follow the procedure and had fun with it, but I didn't really understand it. Then I went to kindergarten. The Montessori way of learning numbers is to represent them by beads for the units, by rods of tens for the tens, by squares of a hundred for the hundreds, and by cubes of thousands for the thousands. You can represent and do math with large numbers this way—but it teaches you why, and you start understanding. So this abstract algorithm that I was being taught by my father suddenly made sense, and I could see that it was correct. Now, this was an overwhelming experience, comparable to other big experiences that you have later in life. So when I wrote my research monograph on lambda calculus, I dedicated it to my parents and to [Maria] Montessori. About Montessori I said, "She taught me with her teaching and introduced me to the experience of truth."

Montessori influenced me, I think, by putting such a strong emphasis on our experience. This educational philosophy also emphasizes training your senses: you have to develop your taste, your touch, and all the senses.

In mathematics such experiences are very important. Most people think that mathematics is based on calculation and on proof, and it is. Behind the calculation and the proofs, however, there is our judgment. The phenomenological judgment that leads us to say, "That is correct," is the real basis of mathematics. This fact makes mathematics very different from the other sciences, and I have tried to show how this links it to Buddhist phenomenology.

So there's a parallel between mathematical judgment and the types of judgment you've employed in Buddhist phenomenology?

Yes, indeed, there is a close parallel, and a few mathematicians have recognized or acknowledged this fact. Husserl was one. [Kurt] Gödel—who realized that mathematics is really a phenomenological science—was another.

About a hundred years ago people in psychology were trying to use introspection as a means of getting psychological information. For many this approach was disappointing—because they realized that many people made unwarranted psychological projections—and they abandoned it. But I think it was too early to abandon the approach. Imagine instead that this introspective approach would be employed by researchers who had developed a trained mind. The researchers would learn to block projections as well as they could, in the same way as mathematicians have to train their mind to do the proofs or physicists have to develop tools to assist in observation. The tools get more refined; first, you have magnifying glasses, then microscopes, then electron microscopes. The same applies to phenomenology. You have to work on it and refine your tools, even if they are internal tools.

When I was six or seven I decided not to become a mathematician. I had asked my father, "What's a mathematician?" He answered that it is somebody who does computations in a refined way and who teaches these skills to other people. I thought, Well, I can imagine doing that, and I can imagine how one becomes better and better at it, but what's the purpose of it all? Isn't it just playing games, so to speak? No, I don't want to do that. But later, when I was twelve, I was exposed to geometry. Quickly, I finished the school material. Then one of our teachers, who was an assistant professor at Utrecht University, taught me some university mathematics.

What was beautiful was the rigor. May I quote [Robert] Musil? Musil said the precision, strength, and certainty of this mathematical thinking, which is unequaled in life, almost pervaded him with melancholy.

Melancholy. It's a romantic term that implies that which transforms or transfixes, that which overcomes one. In German romanticism melancholy has a positive sense too.

Right. Here I think it has this positive sense. But there is also the sense of being overwhelmed by this beauty, by the incredible force of mathematics. In effect, this force is a continuation of this "self-control of error" that was being taught since kindergarten.

At some point I started to reflect about what we were doing in, say, geometry. If people like to solve a puzzle of a certain kind, well, then they can solve it—and some people are happy with that. But other people want to do something different. At some point it went so well with mathematics that I thought, Well, we can continue with this, and it will be always the same—looking for patterns.

In some sense I transcended the doing of mathematics and then I became interested in logic. Logicians, though they come from mathematics, don't do mathematics for the problem solving and the theory building itself but to study the way the reasoning—to study how the mind works, by trying to understand how we are able to find the theorems and to find the patterns.

What you've said sounds like [Gottlob] Frege's metamathematical reflection—in which set theory becomes the foundation for mathematics, in which one moves outside mathematical problem solving to find a logically more fundamental context within which mathematics can be contained and expressed.

Yes, yes! So I myself am much more a metamathematician than a mathematician.

Doesn't that show a philosophical bent of mind?

Yes. Some of my colleagues do the problems themselves, and they do much better work than I on that mathematical level. And I prefer to reflect at the metamathematical level.

Here's a philosophical inclination. When I was fourteen I wondered,

Why do we need axioms? Isn't it disappointing that we can prove everything except the axioms? I wanted to prove them too. Could we not prove them if we could define everything very precisely?

Then another of my teachers made me read Tarski's *Introduction to Logic*. Tarski's answer is simple: One cannot, because if one wants to prove something, one has to prove it from other assumptions. That's all that mathematics is—working from assumptions to conclusions. And if you want in turn to prove those working assumptions, you need to employ other assumptions; hence, as you see, you get an infinite regression. In effect, this consequence was already evident to Aristotle. Aristotle very clearly described the mathematical enterprise, the axiomatic method. He held that in mathematics there are two kinds of things, namely, objects and their properties. Now, how do you get the objects? Well, by definition, from other objects—and in order to have a ground you need primitive objects that are not defined. Then, not long after Aristotle, Euclid wrote his tremendously impressive geometry using this approach of objects and properties. For centuries after Euclid the attitude continued to be that a primitive notion is so clear that it doesn't need a definition, and a primitive theorem [an axiom] is so true that it doesn't need a proof.

We had to wait more than two thousand years for David Hilbert to come up with a completely satisfactory solution to this problem. He said that we don't know what points or lines are in themselves. But all that actually matters is whether the points and lines satisfy the axioms of the system—whether they satisfy the rules of the game, so to speak. So why are there axioms? Well, the axioms are kind of an implicit definition of the primitive notions.

In comparison to earlier attempts, this answer is fully satisfactory. For now we can look at reality and ask, "Where do we find other things that behave like the objects in the axioms?" And then we can apply all the theorems of a given formal system—whether or not the things in the world are the kinds of objects we originally named in the axioms. Perhaps you can see—Hilbert's very general insight is as beautiful as discovering a theorem, such as when you read Pythagoras's theorem.

So at fourteen years of age the young mathematician, reading Tarski and Hilbert, became a metamathematician.

Yes. Later I read the thesis by the American logician [Howard] Goodman, on intuitionist arithmetic. [Intuitionism is a view of mathematics and a critique of classical logic proposed by the Dutchman L. E. J. Brouwer.] Goodman used a technique called the lambda calculus to manipulate proofs. I was so taken by this technique that I studied it, wrote my thesis on it, then wrote a research monograph on it. Each time I learned something new about the lambda calculus, I was again taken by the beauty—whether it was within mathematics itself or whether it was in the reflection. In all cases there was this strong quest for beauty.

Do you have some sense that what's beautiful is more likely to be true?

Yes, and in that sense I like the Platonistic "the good, the beautiful, and the true"—they are just three sides of a coin.

The power of the lambda calculus is this: it's a single language able to describe both mathematical computations—in fact, any computations—and mathematical proofs. Two important results follow for technology. One is that we can use it as the basis of new kinds of programming languages that have an extreme clarity. These so-called functional programming languages are wide spectrum—you can do anything with them, from low-level machine functions to higher-level algorithms to even higher-level interactions with humans. They are modular, meaning that they consist of smaller pieces that go together in an understandable way to make larger pieces. This gives us a grip on what we call the "software crisis," in which nonmodular programs become too complex and many errors creep in. Lambda expressions can be given a type—which is rather like a dimensional unit in physics—and the type is smaller than the expression itself. This allows for error checking—a bit like the Montessori dictum applied to high technology.

As I understand it, the lambda calculus has in fact been employed very fruitfully in computer programming as well as in computational theory.

Yes, both. On the theoretical side, the whole notion of computability that started with [Alonzo] Church was formulated in terms of lambda calculus, and [Alan] Turing later showed that his notion of the Turing machine was related to Church's notion of computability through the lambda calculus. The lambda calculus was there at the birth of computability theory in the thirties.

On the applied side, the LISP programming language was based on the untyped lambda calculus. In the seventies the lambda calculus with types became popular and made "cleaner" programming possible. Then, since the late eighties and early nineties, two new languages named Clean and Haskell were introduced. They are very pure and very powerful.

We have been a little bit disappointed that they didn't catch on world-wide in the software industry. We recognized that the old languages would have to go because their complexity is too great. But then others invented object-oriented languages like Java. These are only halfway functional languages—on the top layer they have a clean structure, but below the top layer they have an old structure, which is more chaotic. Still, because the top layer is so clean, the object-oriented languages are much improved over their predecessors.

And the second technological impact of the lambda calculus?

The second aspect is that with the lambda calculus one is able to describe proofs. This relates to an argument by Brouwer, the intuitionist. Suppose you wonder whether a certain proposition A is true. I say, "For ten dollars I'll give you the answer—in fact, I'll give you a machine that will give you an immediate answer, guaranteed to be correct." So you give me your ten dollars. Then I give you, instead, two machines—one that always says yes and one that always says no. And I say, "Well, it's either true or false. If A is true, this is the machine (the yes machine); but if A is false, that's the machine (the no machine)."

I want my ten dollars back!

So you are an intuitionist. Your response to the scenario is exactly the essence of intuitionism. You could say that classical mathematics is "mathematics on the level of God," because God, knowing everything, knows whether A or not-A is true. But we don't know the answer and, not knowing, we cannot reason from it. Yet as mortals we have to stand on the firm ground of what we actually know. In this spirit Brouwer proposed that we should only treat a statement as true if we have a proof of it, because we are not omniscient and need proof. What is the proof of "A implies B"? Simply put, it means we must have a method whereby, from a proof of A, we can produce a proof of B. Hence we need an algorithm, a function, that converts proofs of A into proofs of B. Now, since lambda calculus describes algorithms, lambda calculus is also good for describing proofs. And what's the technological impact of that? It means that it's now possible to formalize proofs in such complete detail that a computer can verify their correctness. Now we are capable of formalizing any piece of mathematics.

In the seventies, computer scientists were promoting so-called formal methods to prove that their programs were without bugs. The difficulty was that for a ten-line program they needed a thirty-line proof of its correctness. The next question is, Who or what is checking the proofs? If you have a program of a million lines, you need three million lines of proof to be checked. The human mind is easily overwhelmed, so we want the proof checking to be done by another computer program. The Dutch mathematician [Nicolaas Govert] de Bruijn showed that the proof-checking program can be very short, because it only needs to learn the rules of logic. The resulting technology of checking, and hence of verifying the correctness of longer programs, is being used more and more in industry.

Finally, one difficulty still remains: one has to construct the proofs. My main drive now in metamathematical research is to analyze how mathematical intuition works and to teach this to computers. Of course, I don't think we are yet ready to teach a computer what a world-class mathematician knows, but we *can* teach a computer what an undergraduate knows. That level of intuition we claim we can understand; indeed, most of it we have understood well enough to put onto a computer. If we can build a kind of heuristics to do this, then we will see a very different way of doing mathematics.

Let's move to your work in Buddhist phenomenology. What do you mean by this term, and how does your understanding relate to ancient religious or spiritual practices?

I've long been puzzled by the consciousness problem, and at some point I noticed the Buddhists' saying that they understand consciousness—and then I really became interested. I was interested before—a Zen temple looks beautiful and it's a pleasure to sit there and get a calm mind. But when I heard the more traditional Buddhists—namely, the Theravada Buddhists in Sri Lanka, Thailand, and Burma—say they know about the structure of consciousness, I was really driven to study their claims in a deep way. At ten-day and monthlong retreats I learned some remarkable things. One of the key things we learned is a kind of reductionist view of the mind. As I've emphasized in my writing, in the end we need to combine reductionist and holistic approaches. Still, there is a definite reductionist aspect of the mind. Even small children know that, when they repeat a word like *yellow* many times, they suddenly feel as if they don't know the meaning of *yellow* anymore.

The Buddhist understanding of this phenomenon is profound. The Buddhists would say that when you hear a word, there are two aspects, namely, the sound and the meaning. Usually, those two occur on top of each other, but when you repeat them a lot, the sound goes to the left and the meaning goes to the right, as it were. When you look at the sound, the meaning is not there. This splitting into components I call the chemical analysis of the mind. You can do the same with an experience like pain. One can learn to see pain as pure pain, without the discomfort of it. You put the pure experience of it to the left and the discomfort you put to the right. You will see that the discomfort consists of impatience, of wanting to fight, but if you look at the pure pain without the urge to fight, well, then it isn't so painful anymore. Then, when you let go of the fighting, again the pain is a little less and your pain threshold becomes a little higher. All these exercises help you to be able to concentrate better.

However, there is still one thing that is missing. The Buddhists have five *skandas,* or groups. The first is the input, the second is the feeling—the value judgment that we place on the input—the third is the perception or the cognition, and the fourth is the output. Neurophysiologists can work with those four empirically. But the Buddhists have a fifth *skanda,* namely, awareness. You don't only have these four stages, but also you have awareness of them: consciousness.

Once, when I was at Tassajara—this lovely Zen retreat in Big Sur, California—as a student meditator, the head monk said, "Consciousness is suffering." Now, I knew that all life is suffering, but when he said consciousness is suffering, I felt quite down. Consciousness is the most beautiful experience that I have, and he said it causes suffering. Later I understood why: because we're always busy with consciousness. Awareness, the fifth *skanda,* is *consciousness without objects*. We have to let go of the objects and have the pure consciousness.

And if we have consciousness without objects, is it consciousness without suffering?

Yes.

So the monk's message was not that consciousness as such is suffering but only the consciousness that is bound to objects that are ultimately illusory. Is that why you argue for a more holistic approach that includes the fifth skanda?

Yes. I discussed this with [the philosopher] Daniel Dennett when he visited our university, and he said, "No, no, we don't need that fifth one. I can do without it." But I don't agree with him.

That's the difference between a reductionist, like Dennett, and a holist like yourself. For you consciousness is an overarching awareness. It may even be the most fundamental of the five skanda.

Yes, right. In fact, I was a bit attached to it, because one can say that consciousness is my life.

Your work combines meditation, a religious practice familiar to many people, with a thoroughness of analysis and clarity of presentation characteristic of somebody trained in logic and mathematics. It's a very unusual combination and an intriguing one. Do you ever feel conflict as you seek to combine these two different cognitive modes—the releasing that meditation requires and the reflection that analysis involves?

Yes, there is a little conflict, and this is well known at the monastery. For example, we are taught not to think about where our karma comes from. We're also taught not to ask where our consciousness comes from, because we will keep thinking about it and then we won't do our meditation exercises. So every day in the monastery I took fifteen minutes to write down my intellectual thoughts, and then I could let them go. At the end of the meditation retreat I was able to reconstruct the two papers I was working on. The conflict was certainly there, but I resolved it this way.

The Theravada methodology consists of naming. For example, if there is pain you say, "Oh—comma—there is pain—period." By this naming, you let go. Otherwise you're fighting with the pain, but by naming it you put it at a distance. That's the whole point of what's called mindfulness:

distancing oneself from one's pains, from one's thoughts, from one's intellectual drive.

Now, one may also experience a kind of depersonalization in which you lose all firm ground—and that's terrible. I've written that this process of depersonalization is the primary cause of war among humankind. What I really want to do in the future is to find out the neurophysiological basis of depersonalization, so that it might be overcome.

So you think there could be an overlap between the results of the neurosciences and the results of Buddhist phenomenological reflection or meditation?

Yes, and in fact I have obtained money from our university explicitly to do research in this direction. What I call the chemical analysis of our phenomenological mind comes pretty close to what the neurophysiologists do. But it may take a hundred years, so we have to be patient.

Some neuroscientists argue that the brain is nothing more than wires and chemicals and would deny any real causal force to thoughts, consciousness, or awareness. They see an opposition between neuroscientific analysis and what they call folk psychology. Would you resist that opposition?

Well, defining the phenomena of conscious awareness as folk psychology is probably too superficial. But I do think that consciousness has a causal effect. I think that part is very simple.

Other thinkers, especially Western psychologists in the midtwentieth century, would draw sharp distinctions between individual and social psychology on the one hand, and the realm of the spiritual on the other—and they think there's a qualitative leap between psychology and spirituality. Do you allow for such a sharp distinction?

I do not use the word *spiritual* at all here. What I'm dealing with is already spiritual enough. I think consciousness is already special enough, and almost unsolvable enough, that we don't have to add the word *spiritual* to it.

In light of what you've just said, I wonder whether you haven't been overly polite in our conversation. When I've used the word religious, you might consider it inappropriate to your project and practice.

No, because my definition of religion is the road, the path, toward happiness and inner peace. Religion has a common ground with the sciences, since the sciences also require a special state of mind, one that is comparable to the inner peace that religion seeks. Both are driven by the quest for beauty. Moreover, one can study the conditions for happiness and inner peace scientifically.

In that definition of religion, it does not seem that any need, or even any place, remains for the transcendent. Is that what you mean?

Transcendent in the sense of God? Science as it is does not offer evidence of transcendent reality and purpose. The notion of "purpose" is not in science; it comes before science.

Science has not yet understood the essence of the human mind with its self-consciousness, reflection, and purification. It is hoped that it can someday understand the phenomenon of consciousness, which at this point remains metaphysical although known to us all. Perhaps science will reach this understanding through an extension of its methods and domains.

Yet at the moment consciousness is still a transcendent thing, because it hasn't yet been understood in physics or in neurophysiology. If in the future we acquire evidence of what is now transcendent, from that moment on it will no longer be transcendent. Thus I do find a place for transcendence; it can be found in the consciousness that you and I and everybody know very well in our daily life.

Some people believe that consciousness is inherently self-transcending, because a self-referential dimension is involved whenever one is conscious of something. Is that close to your position?

Yes. Humans especially have this dimension of self-referentiality, but probably dogs have it to a lesser degree. But, as I said, I don't fully understand consciousness yet, so I'm not sure what will happen in nirvana when we lose all objects.

In one of your papers you speak of a "process of purification." This makes your phenomenology different from Husserl's, which is merely a chronicling, a science of the inner, or eidetic, world. In your writing the process of purification seems to be an ethical process. Have I understood you correctly?

It does indeed have ethical implications. When in the process of purification we see that we have forces in ourselves that are capable of harming other people, we realize that we have to overcome them—because we know the forces are in us and not in the others. Second, when we see that other people are still struggling with their own harmful forces, then we have compassion for them. We see that they have to hide their struggle, that they are busy treating their process of purification in a symptomatic fashion, and that the treatment has become a kind of addiction. That recognition gives great compassion.

The process is one of the purification of defilements. The five *skanda* are bound up like a big knot, and the knot has too strong a grip on us. We have to free ourselves from it. What we have to gain, though, is a bigger space of choices. It's as if we were insects: a moth will fly into the flame, but we must go higher so we can circle around it without burning ourselves.

In your writings you use the phrase "mysticism and beyond." Among students of the world's religions, mysticism has traditionally been understood as the tendency within the religious traditions to move toward that which is absolutely unitary, which is the highest object of study or knowledge, and which moves beyond all words and all expression. Do you mean mysticism in that very strong sense?

I do mean it very much in that sense, and I've described it as a state of sublime consciousness. But at the moment [that] you think it is the highest state you can reach, the teacher comes in and says, "No, you also have to detach yourself from that, and you have to name it."

What you've just said stands in sharp contrast with some of the Christian mystical traditions, where this state is called oneness with God and is treated as that which is most to be sought after. But in the Buddhist tradition it is not to be sought after; instead, even it is left behind in the end.

In this sense I see Hinduism as more comparable to Christianity; "Atman is Brahman" is comparable to Christian mysticism. Yet even St. John of the Cross warns his readers about wanting that state of God too much—so much that we get addicted to it.

In Buddhism nirvana is a positive state. *Nirvana* means "extinguished," but this refers to the extinguishing of defilements. The Buddhist view goes further, to a final purification. That's why I call it mysticism and beyond.

Let's return in closing to the question of integration. It seems that our discussion has had two very distinct parts: your work on the lambda calculus and its applications, and your work in Buddhist meditation and phenomenology. Are they for you distinct, even opposed in the end, or do you find some sort of integration between them?

I would say both. Science and the spiritual quest are as distinct as Mozart and Beethoven, but they are the same because both point toward beauty. In the scientific quest beauty of understanding is the driving factor. Scientific understanding is explaining complex phenomena in terms of simpler ones—and in that sense science is reductionist. The spiritual quest, though, should be both holistic and reductionist.

There should not be any tension between science and religion if both are done well. For this to occur, scientists have to expand their horizons and religious leaders have to relativize some claims. The fields of science and religion are not inherently closed to one another, though some *people* may be. I do recognize religion based on faith but prefer for myself another form of religion.

As in the case of Mozart and Beethoven, you need similar mind-sets and a similar calm. You need a purified and trained mind to do mathematics, and you need a purified and trained mind to do meditation. When you have both, things fit together in a different way than you saw them before. This is insight—an important word in mathematics and the very meaning of *vipassana.*

3. Khalil Chamcham

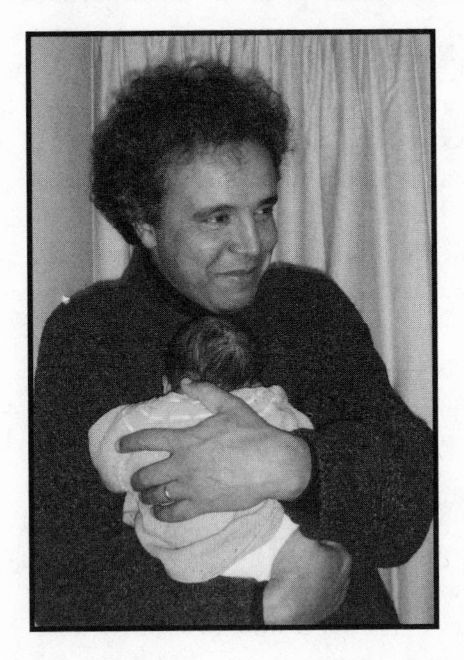

Dr. Khalil Chamcham lives in Casablanca, Morocco. He received his doctorate in nuclear physics from the University Claude Bernard, Lyon, France, in 1983. Since 1988 he has organized Morocco's national meetings in astronomy at the University of Casablanca. He received his second Ph.D. from the University of Sussex in astrophysics in 1995 and has also carried out research in astrophysics and theology at Oxford University. Now a full professor at the University Hassan II–Ain Chock in Casablanca, he has initiated, for the first time in Morocco, undergraduate and postgraduate curricula in astronomy and astrophysics. His research interests include the stability of galactic discs, star formation,

---

Philip Clayton interviewed Khalil Chamcham, and Holly Vande Wall and Zach Simpson edited the interview for publication; it was transcribed by Jessica Hazlewood.

and the chemical and photometric evolution of galaxies. He is a member of the International Astronomical Union, where he is involved with the Commission for Education and Astronomy Development.

A practicing Muslim, Chamcham has helped to organize international colloquia on Islam and science, including a colloquium on the history of Islamic astronomy at the Osservatorio Astronomico di Roma in Italy. In addition to studying scientific development in the Islamic world, he is a student of comparative religious thought, including advanced studies of Christology in the Christian tradition.

Khalil Chamcham speaks an elegant French: sophisticated, literary, reflective. But even when discussing his views in a more pedestrian language such as English, his speculative skills are evident. Deeply philosophical by nature, his mind moves instinctually and continually to the borderline areas between science and religion; he stares off into space as he struggles to find ways to express their interrelatedness and compatibility. Watching his deep, restless eyes as he talks, one senses a mind equally fascinated by the puzzles of galaxy formation and by the questions of how to relate divergent religious traditions.

Welcome, Khalil. Perhaps the easiest way to introduce yourself would be to describe the development of your work in physics, beginning with your graduate studies.

Thank you for having me. I suppose I would describe my interests in physics as a continuous evolution. I received my doctorate in France, working primarily in nuclear physics. And then, after receiving my doctorate, I moved into nuclear astrophysics and have subsequently developed a parallel interest in the chemical evolution of galaxies. I have since done a lot of work on the formation of galaxies, with my concentration being especially on star formation.

Which research groups have you been involved with to explore these areas of interest?

I began my work at the Institut de Physique Nucléaire at Université Claude Bernard in Lyon. I then moved and did my primary work on nuclear astrophysics at the Institut d'Astrophysique de Paris. I finally moved to the University of Sussex to start my research in astrophysics but quickly moved from there to Glasgow in order to pursue a project I have been developing there since the time of my arrival.

Can you explain some of your current projects?

My current work in Glasgow is to develop a photometric model of galactic discs and to attempt to put such a model in a cosmological context. We are trying to formulate some of the constraints on cosmological parameters.

I have also worked on another project, involving the chemical evolution of galaxies, with Professor Francesca Mateucci at Trieste Observatory. My models of the relationship between star formation and the chemical evolution of galaxies are basically rooted in the work of R. [Robert] Kennicutt. He produced a series of observations in 1989 showing that the star formation process is related to the dynamical properties of discs. I am employing his results in an attempt to build a more consistent theory of star and galaxy formation, where I show that galaxies start forming from a given initial time, and star formation starts only after a delay of a few gigayears. I am modeling this discovery to see how it affects the predictions from the previous standard models, in which star formation began simultaneously with the evolution of discs.

Using a group of other findings, I am also proposing a theory of discontinuous or intermittent star formation, which implies that star formation is not a continuous process, but is, rather, a discontinuous one. This theory would help explain some observations we've had of abnormal metal "dispersions," or distributions, on receptor chemical elements that we've analyzed.

These dispersions have to do with the time metric's not fitting, right?

Yes. It is related to the time metric, or at least to the correlation between time and metal distribution across these receptor elements. When you observe the correlation between age and chemical or metal abundance in our galaxy, for instance, and especially in the local disc, there is significant dispersion. Most theories predict a continuous correlation between the age of the galaxy and metal dispersion patterns, but the most updated data show that there is no correlation. In my point of view, the lack of correlation is related to intermittent star formation, in that, if star formation is discontinuous, if it goes in and out, then you destroy any time correlation between metal distribution and age.

Is there any theoretical structure to understand how star formation might be intermittent, what that might mean? That's a revolutionary idea, isn't it?

Well, I just work on things, and it's up to the community to say whether that's revolutionary or not. [*Laughs*]

That's a good cautious physicist speaking.

Yes. In theory, star formation has an ability to stop itself, as it's a self-regulating process. However, once star formation is occurring, the evolution of new stars may slow down, as the interstellar medium may not contract to form stars for a long period. There is a feedback mechanism that reciprocally relates the formation of stars and the interstellar medium, making them dependent on each other. Star formation cannot occur without a contraction of the interstellar medium, which can be quite discontinuous. So if star formation is also related to the interstellar medium, which is itself discontinuous, one would expect more discontinuous patterns of metals distributions through time in the galaxy.

You also do work in magnetic fields, right? What effect do these have on star formation?

It seems that magnetic fields may have some effect on star formation, but it is unclear what role they may play. The clouds out of which stars form are magnetic, and they always remain in that particular magnetic field. I am researching how these magnetic clouds influence the star formation process, but it's not easy to understand how the magnetic field inside the galactic discs stabilizes the disc and then delays star formation, given that we know very little about the role of the magnetic field in star formation itself. So a lot of work has to be done to determine the role of these magnetic clouds in star formation, before we can even begin to assess how they help regulate discontinuous formational processes.

What impact on the field of star formation do you predict the new data from Hubble and the upcoming European space missions will have? Do you expect a radical impact on your field?

I think the Hubble Space Telescope has given us very rich data to use in understanding the process of star formation, especially in understanding formation processes at high redshift. [*Redshift* refers to the spectrographic analysis of light coming to us from distant stars. That it has shifted toward the red end of the spectrum shows that these stars are receding from us at increasingly high speeds.] Hubble, and many other instruments like it, can give us clues as to how efficient galaxy formation is at high redshift and, indirectly, what occurs in local galaxies at lower redshift. Because of Hubble's ability to see into distant galaxies—which means much younger galaxies—it can give us significantly more clues as to what occurred early in galaxy and star formation and, specifically for me, how star formation occurs over time.

The redshift question links your specialty of star formation with some broader cosmological questions, like the formation of the universe, doesn't it?

I think Hubble may allow us to answer a whole suite of questions, like whether we are currently in a hot or cold universe or even the values of certain cosmological constants. Unlike many astrophysicists, I emphasize the question of cosmological constants, because I come from nuclear physics and have a background in quantum physics, where it's natural to think that space is not a vacuum, a nothingness, and therefore it should have some properties that explain its behavior. And these are expressed in constants. We may correlate, for instance, a cosmological constant with the energy content of the vacuum, as it's shown in some models. Such a constant could have tremendous explanatory power. My understanding of physics is such that I've always believed there to be a cosmological constant. It just needs to be discovered.

As for the redshift, I think the main issue on which I have been focusing my attention is the concept of time. The further we regress into the redshift, the more we reach back to the early history of the universe. For many reasons the concept of time has been somewhat neglected in cosmology. But, needless to say, this is an important area of inquiry, because it's related to the question of origin, of, potentially, our origins. So as long as we don't understand the concept of time, I don't think we can give any proper answer to the question of origin. This is where I believe the exploration of deep space to be pivotal. It has the potential to unlock many of our most pressing questions regarding time and the origin of space.

Are you expecting more problems for the general theory of relativity, then? Is it a radically new conceptual structure that you may be forced to accept in the end? Is the current notion of time inadequate?

I think the general theory of relativity has to invest more effort in the concept of time. I think from the conceptual point of view, cosmology needs more understanding of the concept of time. This is my personal speculation, because right now we are usually using it as a free parameter to describe the evolution of phenomena or just to describe the fourth dimension of space. I think there is more behind the concept of time embodied in the structure of the universe, in which case we have to find out what is behind this concept and potentially where it comes from. I think in this regard the general theory of relativity may not be completely adequate to address the fundamental question of time. It likely needs to be supplemented with a more current notion of how time operates under high redshift or a notion of time grounded in the origin and formation of stars and galaxies.

I'm struck at this point in our conversation with how detailed theoretical and experimental, or observational, work on star formation leads in such a natural way to cosmological questions and from there to very fundamental questions that move beyond what's currently testable but seems so crucial for formulating adequate models. I think, for instance, of the anthropic principle itself. [A principle or argument in cosmology is anthropic when it uses the actual existence of humans to help support conclusions about the nature of the universe or about its past history.] An outsider to physics is struck with how philosophical some of the major debates within astrophysics seem to be. Is that your impression?

I think those more general philosophical debates are necessary. Especially notions like the anthropic principle that continually remind us, as scientists, that science is the product of human beings, that it is a human endeavor. Because I can't divorce science and the human pursuit of models and understanding, I think we have to employ our own understanding of time to understand the formation of the universe, how it operates. In this way I have to correlate my concept of time and the anthropic principle. Otherwise, I can't really articulate the relation between our understanding of time and how the universe operates. You can also put it in the form of a question: If the universe already has time in it, why do we need time, as a human concept, in order to describe it?

Yes, that's an excellent question. Can you speculate on what response you would get?

I think a reasonable response would be to say that if we are the intelligence of the universe, we are also the product of the evolution of the universe. Time in that sense might be our need to understand the evolution of phenomena and maybe even ourselves. Hence what we describe as time is just the projection of our fundamental need to understand that we are locked in something called time, and we then project this particular structure, time, on the universe. This is how I think the concept of time and the anthropic principle should be related. In fact, the notion of time may be the best example possible of the anthropic principle. In the instance of time, they cannot be disentangled. I think that as a consciousness, especially an evolved one, we are locked into chronological events, a biological clock, because we need to understand that we have an origin, and we need to understand what death is. Why do we die? I think our own death forces us to consider time more forcefully. And the concept we happen to attach to the evolution of things, to satisfy ourselves, is this thing called time. But that still doesn't answer the question, does it? We're still left wondering whether the universe as a whole *needs* this thing we call time.

Yes, very interesting! Your question suggests a related question. Is consciousness at home in the universe or does it have ideas, intimations, questions that find no parallel within the physical universe? The worry would be that we raise questions about the importance of time and place value on the immortality of the soul in ways that might be, under a naturalist worldview, utterly foreign to the world that we live in. I guess I'm wondering about a correspondence between our intuitions about the world and the world itself.

I think this is why the question of time is so relevant. Because, if we're historically conscious beings with an evolved consciousness of time, then why have we evolved a coherent and guiding concept of time? Doesn't that imply that there's some parity between my concept of time and that of the world? I guess I prefer to simply leave this as a question. As I've said, it's all related to the question of origin. You made the point that if this is an upper consciousness, which is running this thing that we call time, that is a reason why the question of time needs more investigation. And I think our answers to this question are vital, not just from the scientific point of view but also for metaphysical or religious reasons. Resolving this question, or the ones like it, may even lead to new scientific developments in the field, by letting ourselves be open to what is not physical. Or at least to new concepts of what constitutes the physical.

This issue leads to the heart of a whole series of questions that cosmology inevitably begs. For those of us who believe in a god, is there any evidence that we can find within the universe a concept that points toward the reality of time and a universe that might be a creation of a very conscious and very intelligent being or, in the end, is that merely a matter of our faith, completely separate from our study of the natural world?

I think asking this question could make some progress in our ability to account for time. If we say time is only a reflection of our conscious need as human beings, because we are the product of an evolutionary history, even an infinite consciousness, we are then locked inside, and mirroring, this clock that made us. Therefore we need to, and inevitably will, in facing the infinite universe, project our own categories of time, correlate ourselves physically to this universe. But we have to ask ourselves: Is this simply some kind of psychological projection [that is] merely a product of evolution, and therefore time, not something intrinsic to the universe? Or, alternatively, if time is there and we are part of something that is connected to the universe, is there an upper consciousness, something like God, that is there to generate this thing of time?

We still don't understand what time is but can describe events and put them in chronological order. If we say, "This is how we describe them in time and this is what time is," aren't we confusing time with history?

Dr. Chamcham, would you talk about the way that you understand this transition from physical or scientific questions to questions that are really religious questions? Is it a gradual transition or is it a leap of faith? When a human being makes that transition, what occurs?

I should probably first clarify what I mean when I say religion. For me, religion is not merely what I perform in ritual or in a church or otherwise, but it is the aura that I live in, the spiritual force that makes me open to different ways of understanding. Religion allows me to encounter and explore unexpected or unexplored ideas. And I believe this confidence, the confidence that religion gives me, is necessary to head into new ideas. The future is not always clear for us, but I don't relate it to practice, I understand science as a religion somehow. It's a practice, a religious practice, because it's the way for me to show my humanity.

If there's any upper truth, I mean a universal truth, science is a way that may lead to this truth or show fragments of this truth. We are human, so I don't think that science will be something so powerful that it will explain everything that we desire to understand. We are finite, we are not able to control everything. Science is the product of our intelligence and will remain in our image. I think the human side of science shows itself through this incompleteness. As long as science is incomplete, it is human, and it will always remain incomplete. This is why being open to other questions, maybe even religious ones, is not in conflict with science.

So it's not as if there's scientific knowledge, and then the human religious or spiritual side is separate from that, but, rather, there is for you a smooth continuum between science and religion? We can attempt to learn what we can learn from science, we can ask broader questions that perhaps our science cannot answer, and we formulate some sort of response? For instance, the question: Is our existence meaningful in this universe or is it meaningless? That's not a question that science can answer, but for you it's a question that somehow is raised in and through scientific work?

I would say that I am a scientist, and this means to me that in every bit of progress I make in my research, my understanding continues to be enlightened. And the better I understand what is going on, the better I understand myself. Each conclusion, or finding, always leads to a quest for more answers. And I think that's all you can reasonably expect. And, because I'm human, my weaknesses are always in my work. Science is a way, though, to justify myself as a witness and to give more support and more understanding to what I am witnessing in the universe around me. I don't like asking, Is our existence meaningful or meaningless? We are here, and the fact that we are here is some kind of glory, some kind of happiness and joy by itself. Preserving life and hoping for more understanding and living the excitement of discovering and understanding things as a human community of scientists—this is some kind of truth in itself, one that I think is conducive to a religious way of living as well. In this way some of the questions we've been discussing naturally lead to the religious, or a religious way of being and living in the world, in a certain sense. The scientific quest in itself can be a religious way of being through the questions that it raises, through the sense of discovery that it creates.

So it's not as if there's an objective fact of the matter. We make human existence meaningful by how we exist, by how we live?

Precisely. I think that even in the Bible or Koran, God had to do some work to create this world. I think work, intellectual or physical, is a way to make life meaningful. We are responsible for our life, and that responsibility always implies giving meaning to our lives. And, in a particular way, that's the function of work. Every person who works is giving sense to his or her life and his or her community. This is meaningful enough.

It makes me wonder what sort of responses you have, as you do your work, to the incredible beauty of the photographs or the theories that you work with. Is there a response that goes beyond the purely rational scientist when you work on your area, star formation, a response that you might even speak of as religious?

I'm always struck by the beauty of a theory, of a natural phenomenon, or even of the harmony I seem to find in things. And I think this sense of beauty and awe is missed by science. Science may have the progress of medicine and so forth to relieve human beings from suffering, and it may give some happiness, but there is something that cannot be sorted out by science, namely, a sense of wonder, your emotions, the sensation of something wonderful. I think when, as you said, you look at a picture and you see the beauty of the world and purity of life, you're struck by the beauty first. You cannot put that into an equation. You are a human being and you have feelings that will never be controlled. I am interested in artificial intelligence, which attempts to explain such sensations or feelings, but our control or understanding of these things cannot prevent you from having this astonishment or joy when you look at these pictures or look through a telescope and see the beauty of the universe. It's not only science: it's beauty and harmony, and this dimension is a continual stimulus to do scientific work. This is what keeps me happy in doing my research—the fact that I am able to frequently encounter beauty so intimately. My work often is worthwhile for the sake of the pure beauty that I am able to witness.

So there are responses that the scientist has as a human being that could never be contained within equations, and these responses of beauty or awe or even reverence are important in that they motivate the scientist in doing his work?

I think they are. People don't have to be ashamed to acknowledge that their motivation is from the aesthetic—or worried that people would point to them and say they are not pure scientists. I don't like the idea of being pure. [*Both laugh.*]

In an indirect way I think the aesthetic does serve as a motivation to pursue the unknown. We typically have to work hard on the technical side in order to make progress, so we need some inspiration. The inspiration may come from wonder and contemplation. The sense of wonder can come from beauty or a feeling of harmony or even pure awe. And I think contemplation is something very important to make an opening to the unexpected or to discover the unexpected. I feel that contemplation is a very important practice in daily life as a way of developing my inner harmony and to feel the harmony of the universe.

Even from a scientific perspective, though, the most beautiful equations are often the ones that also have the greatest powers of explanation. In aesthetic terms you look at Einstein's or [Werner] Heisenberg's equations—they are so beautiful, so simple. But once we try to go through the simulations, we find that they are so complicated. But finding such an elegant solution is, in itself, a motivation to further discovery.

Interesting. When you use the word contemplation it reminds me of the word meditation in the Eastern traditions, like Hinduism and Buddhism. Is there something similar between what you mean and a religious understanding of contemplation?

Certainly. When I think of contemplation, I am often thinking explicitly of Muslim Sufis, who often meet high levels of inspiration through the practice of contemplation. And I think it is this form of contemplation, one that simply allows events to happen, that has allowed the Sufis to attain one of the highest levels of understanding among religious people. And, in a certain sense, scientists also have the opportunity to experience this form of contemplation, the understanding that it has given the Sufis. By making themselves infinitely open to the beauty of the world in contemplation, they are able to understand what's going on much more than ordinary people who are just technical or mechanical in their work. In being contemplative, in allowing the world to open itself up, scientists have essentially opened themselves up to this religious dimension of understanding.

Is there a type of knowledge that goes beyond the quantitative knowledge of the equations? Is this thing that we've been talking about with contemplation purely emotional, or is it a way of knowing?

I think it is both, because the form of contemplation I envision opens up avenues of understanding that are both spiritual and scientific. It's not really a matter of scientific or emotional understanding but more a matter of going as far as possible within your own understanding of the world, of your ability to explain things. And these dimensions can be both religious and scientific. A side question to all this, of course, is: By contemplation do the phenomena become more open to you, do they really reveal, or did we just discover something that didn't exist at all? It's really a question of whether one discovers or one invents. And the idea and practice of contemplation is more closely related to discovery.

Well, classically, the religious people, Jews and Muslims and others, have said they discover a revelation of God, but I'm not sure under your perspective if you would want to use that sort of language.

I don't think I would use that language simply because I'm cautious—I don't feel I can understand God, much less God's secrets. What would it mean to discover a revelation of God? If God has created me as a witness, I still have to understand what I'm witnessing and who he is, such that he would make me a witness. Maybe through contemplation you reveal what is already there, as things and events show their secrets because they open themselves to you. It's related to the harmony that is there already—maybe this is the role of contemplation. To contemplate what it means to be a witness, to be in the presence of God. Maybe the three Abrahamic religions are right when they give the answer that this contemplation is the revelation of God. Or contemplation may simply be a revelation, not necessarily religious, that reveals what is already here, the harmony already present. And science can be a particular form of this contemplation, one where you allow things to reveal themselves to you because you have put yourself in the state where you could see this revelation.

Even if we don't want to use the word knowledge about this phenomenon of contemplation, it sounds as if you would describe it as a deeper level of understanding or comprehension that's available to humans, that's not purely scientific but somehow transcends science.

I think that's an excellent description. And, like I said before, so much of science, especially good science, depends on this level of comprehension and contemplation that often transcends the strictly empirical.

Interestingly, scientists cannot be happy unless they achieve this level of comprehension, that ability to contemplate and think in ways that reach to the depth of their experience. This is most certainly the case with great scientists, who have often reached deep into the phenomena of the world to make incredible discoveries. And so I think we have to learn about the personality of the high scientists who have revolutionized our understanding and which spiritual state they were in when they made the discovery. I suspect that their level of understanding was tantamount to that of religious contemplation.

I'm very interested in the harmonious relationship that you describe, between the human scientific quest and this spiritual quest. You clearly don't identify the two, but you also don't describe them as two things that are in tension or at battle with each other. Rather, there's a sense of their being two aspects of the human person or of human personality. Or possibly of the same way of being. Have I understood correctly?

You are correct to see that I don't disentangle scientific prospects from the human experience; science is our product, it's not something that is independent of who we are. The same holds for religion. I think we as humans are a transition, a tightrope, between the universe and who we are, always in a position of attempting to discover our harmony within that universe. Science and religion are both manifestations of this need to translate our experience of the world into, and in, our own terms. So I don't agree at all with people who think that there is only scientific understanding, and scientific explanations, of the world or only an absolutist religious understanding of the world. It's one single movement, and we cannot disentangle that from life. I think that we cannot disentangle the two processes, scientific knowledge or spiritual experience, from human beings.

I asked at the beginning of our discussion about the development of your scientific specialization and your viewpoint. Could I have your permission to ask now about the development of this religious or spiritual viewpoint? In other words, what have been some of the influences on you? Did you came from a very religious background, or did these ideas arise from your own reading in philosophy and so forth?

I must say that my parents have been very special to me in that they always left me free to choose—they never obliged me to believe or to practice anything in particular. I am from a Muslim background, and my parents are very religious, but they have a deep respect for who I am and who my brothers and sisters are. They never felt that they should oblige us to do something; we were always allowed to do what we wanted, and they had faith that we would choose wisely. Thanks to this way of parenting, I've been allowed to cultivate my own attempts to understand the world and to continue questioning. As I've grown older, these interests have only grown. It's partly from my religion and my discussions, and partly from my experience, because when you travel a lot, you meet so many people of other cultures and religions, and, if you remain open to their world and experiences, much can be learned.

I think that it would be a big mistake to try to lock myself in one single religion. As a Muslim I love so much the faith of Christ, because it represents so many values for me. And by saying this I don't cease to be a Muslim. I'm simply acknowledging some truth or aspect in my particular human dimension, the dimension of Christ. There is a lot to learn from Buddhism and other religions as well, but I don't practice those religions. Because I can see the religious dimension expressed in so many ways, I don't ever allow myself to be confined to a church or synagogue or otherwise, or to prevent myself from attending those forms of worship. I always find a way to meet other human beings who are walking on another track and to tell them, "Let's walk on the same track, let's not be so disconnected from each other." Of course, some people are limited and stay in his or her own box. But occasionally I am able to walk with others, see their path, their track. And I think I reached some spiritual or natural evolution of my personality because I have been taught to do this from the beginning.

It's interesting to me that you still describe yourself as a Muslim, as being a practitioner of one religion, and yet you do so with a great openness to the results of science, to other cultures, and to the insights of other religions and philosophies. That openness of mind-set, within your self-understanding as a Muslim, is very intriguing.

Yes, I think that's an adequate description. To use an analogy from star formation: star formation is a self-regulated process in much the same way that human beings are self-regulated with respect to intelligence and consciousness and so forth. I think what makes me human is the ability to analyze and discern myself and discover what is wrong and right within me and what is good or bad. This process is often discontinuous, but it does appear to have some telos. In doing this I think I'm ultimately expressing a sense of what it means to be human, to be self-conscious, and trying to meet every human value in every other person. From this point of view I cannot judge any person because he or she believes in something—that's something universal, something tied to human freedom, to our own individual processes of formation and regulation. We all share this common ability and desire to believe, to express our sense of freedom, and I attempt to meet this level in those I come in contact with.

I think there is a clue here, in my understanding of human freedom, in how we understand the universe. I think we can and should ask ourselves, are the laws of physics an expression of some freedom inherent within the evolution of the universe and is our ability to discover these laws, or to turn a blind eye to them, an expression of that same freedom? I think here that the concept of freedom can be universalized, that our sense of freedom manifested in the religions can be understood as stemming from the same freedoms that allowed the universe to evolve. And for this reason it's untenable to reject, or even gloss over, the beliefs of others. Their beliefs are natural, they spring from the same source as the rest of nature. *Not* to meet a human being on that level would be to reject that freedom.

We've talked about freedom and time, and we've talked about this deeper level of comprehension that goes beyond science. But we haven't yet talked about ethics or values. It's often said that science is value free and if there is to be a sense of values, or right and wrong, it has to come from outside science. Do you think that's true, and, if so, where would we get our understanding of these matters of value or right and wrong in science?

I think some great scientists have been immoral by rejecting the role of values in science. Values, ethics, should never be forgotten. Science, and technology, which is ultimately the result of scientific development, cannot turn a blind eye to atrocities in the world—whether they are religious or racial and whether they are minor or rise to the level of genocide. Turning a blind eye in some sense is the same as complicity.

So science strongly needs some values from outside it to protect it against extremes. Is this what you mean?

The values should stem from the scientists themselves. In this way the values aren't really from "the outside," necessarily, though they may not be inherent within scientific work itself. We should always work with ethical consequences in mind and not just let humanity suffer from what is going on, just because we think that the only important things are the results of our work, that progress must continue despite the rest of humankind. This ends-oriented way of thinking can be disastrous if it turns away from the ethical challenges that it raises.

Yes. That's nicely put. I want to ask you, where do you think these values will come from? If they're not given by the science, and some scientists seem to ignore them, where does humanity obtain this sense of ethics? Is it discovered or is it constructed?

These values have to be constructed. In fact, they are constructed. Self-consciousness reveals this. As a conscious being you are always building values, always contributing to the beauty that *is* humanity, always projecting values onto life. Or maybe simply emulating values that are already there. True action simply means responding to these values, to your own established sense of harmony with the beauty of the world. In this way I don't think the values established "outside" science exclude those "in" science. One does not exclude the other. What is revealed to you outside science doesn't exclude what you discover in your own work and your own construction as a person in a group or as a group of individuals working for the sake of the community. Human value creation seems much more unified than that.

Your comments on this subject remind me of what you said about consciousness a few minutes ago. We wonder whether consciousness and time are part of the physical universe themselves or if they are something radically different. I also wonder whether our concern with right and wrong is something that has anything to do with this physical universe that we have arisen out of. Most religions say that this moral sense does emerge from the universe, because it comes from a god. But if one doesn't begin with the belief in a god, it's difficult to understand what to say about our own sense of right and wrong.

Absolutely. I think when you connect this idea of value creation to freedom and knowledge, you could say that we have been created by God and all that we do is created by his majesty. But saying from the start that everything is explained by God might somehow prevent ourselves from discovering more. To say that this is sacred truth, that things are already explained, and we don't have to work for them, or to discover them, is a way of guarding against new discoveries, new ways of envisioning the world. I think if God created us in his image, we are inherently incomplete, and we are unable to understand all the secrets, his secrets. If we say that things can be revealed partially to us, especially if we work for them, there will be more revelations and more clarity. To continually understand that you don't carry the absolute knowledge and the absolute truth, and neither do your neighbors, is to say you still have something to achieve together, and *I* still have something to achieve in my life, and the future generations still have something to achieve for themselves. I think that leaving some questions open, and even leaving some answers open to questioning, is not a manifestation of disbelief. It's a recognition of humility. If we are created in God's image, it is this process of discovery and creation that allows us to improve in his image, to grow closer to that ideal. Value creation is merely a product of this continual attempt, finite as it is, to create and discover.

As we come close to the end of our time, I'm struck with how the last words that you said very much reveal the spirit of a professional scientist, because you avoid absolutes. You avoid claims about certainty, and you want the most important things, the things about our consciousness and our values and our understanding of the world, to be open. It's as if you're saying they should remain hypothetical, like a scientific hypothesis, is that right? Even the area of human religious or spiritual beliefs and practices?

Even if there are some solely religious realities or beliefs and people need some solid statement, I think religions should keep themselves open to questioning, open to more truths. If they say that "we have all the truth because God says everything and I believe in what God says," they close their hearts; when they do that, they become like stone.

That's beautifully described. So true religion in the deeper sense is open to new information, it evolves, it never closes itself to the world?

And religion, or truth, doesn't stop with the disappearance of perfection. I mean the truth did not stop because Jesus is gone or because Muhammad or Abraham are gone—they revealed truths about life and duty, and every human being has a duty to carry on the principles of what they came for and what they revealed. I'm not equivocating here. [*He pauses and reflects deeply, then proceeds slowly.*] Rather, I think our duty is always to understand ourselves, to ask what is the reason of our being, and, possibly, to see where we stand in relation to the universe. Even if we don't want to be stuck in the metaphysical question, to discover the beauty of the universe and transmit this knowledge from generation to generation is a duty and a way of expressing what all these prophets originally came for. So, even if we can't discover the meaning of our relation to the universe ourselves, we can always rely on past generations, on the wisdom that came before us. This is why I don't think religion is something that should be excluded from the life of a scientist. Religion isn't just a way of revealing, it's also a way of passing down that revelation from generation to generation. And science simply can't dismiss this way of knowing.

Throughout the interview you have expressed a sense of freedom. The scientist is free to ask these broader questions and even compelled as a human being to ask these broader questions. And, at the same time, you've encouraged a religious or spiritual response that is free and humble and open to insights from science and experience.

If I have to draw a scheme of the relation between the religion and science, it would be in terms of horizontal and vertical forms of understanding. Religious people have an out/in view of the universe and view life progressively from a vertical to a horizontal understanding. Scientists have an in/out understanding, from horizontal to vertical. Both are one single path but taken from different ends. This is why I don't think there is an exclusion between the two. If we exclude this freedom from understanding, or even belonging to a religion, we end up with exclusion or even genocide—we have to kill the one who doesn't look like me or doesn't belong to my religion. I think a dialogue between science and religion begins with this understanding that we're all on the same path, however opposing it may seem at times.

4. Donna Auguste

Donna Auguste is a remarkable computer scientist, engineer, entrepreneur, and humanitarian. Pursuing a childhood fascination with technology, she earned her undergraduate degree in computer science and electrical engineering in 1980 at the University of California–Berkeley, where she served as president of the Black Engineering and Science Students Association. She completed a master's degree in computer science at Carnegie-Mellon University, where she was one of the first African American graduate students in her department. Her early work at IntelliCorp focused on pioneering commercial applications of artificial intelligence technology. At Apple Computer she led the project team for the Apple Newton, the world's first

---

Philip Clayton interviewed Donna Auguste; Jim Schaal edited the interview for publication.

personal digital assistant. As a senior director at US West she led an engineering team that developed a fiber-optic broadband network.

In 1996 Auguste fulfilled a long-standing entrepreneurial dream by founding Freshwater Software in Boulder, Colorado, which grew into a highly successful business that provides services and Internet solutions for e-commerce. Under her leadership as president and chief executive officer, the company exemplified a highly collaborative model of teamwork. While at Freshwater she also edited the business newsletter *WomenCompute.com* and launched an initiative to bring solar-powered electricity and Internet access to schools and hospital clinics in remote villages in Africa. When Mercury Interactive acquired Freshwater in May 2001, the sale of the company allowed Auguste to embark on yet another career—philanthropy.

Auguste founded and endowed the Leave a Little Room Foundation, a nonprofit whose mission is "to be of service in sharing gifts of gospel music, education, food, clothing, housing, technology, and all resources that God puts into our care." The foundation continues Auguste's efforts to bring access to technology and health care to rural communities in Kenya, Tanzania, and Eritrea. It also builds houses for low-income families in Mexico and provides clothing, textbooks, and computer instruction in underserved school districts in Colorado, California, Illinois, and Mississippi. The foundation's outreach ministries, funded in part by revenue from the digital distribution of gospel music through its LLR Gospel Music division, are described as follows: "Everything that we do is for the glory and honor of God. By sharing these gifts we leave a little room in our hearts and lives to be enriched by The Holy Spirit."

"As scientists, engineers, innovators, and entrepreneurs, we are called to explore," she said. "We are called to step out on faith to seek new understanding. We're called to do tasks that have never ever been done before. We're called, time and time again, to tackle new challenges."

Auguste's commitments grow naturally out of her life experience. She grew up in Berkeley, with close ties to her extended family in Louisiana. Her career-long support for cultural diversity has deep roots: her family is of Creole descent, with ties to both the African American community and the Native American experience of the Choctaw and Blackfoot nations. A devout Baptist, she has been a church musician since 1984; she plays bass guitar for the gospel choirs and codirects the children's choir of Cure d'Ars Catholic Church in Denver.

Can you start us off with a personal narrative? In a remarkably short period you moved from being an undergraduate in electrical engineering at the University of California–Berkeley to team leader of a historic project at Apple and then on to become CEO and cofounder of Freshwater Software. That suggests enormous personal development. Or is it the same person and the same priorities all the way through?

I think in a lot of ways it's the same person and the same priorities all the way through. But some of the most important lessons started emerging for me in college. My college experience was in some ways eye opening and in some ways painful. While UC–Berkeley is known for its liberal side, especially in the '60s and '70s, in the College of Engineering I found an extremely conservative environment. There were faculty who did not want students of color, and they felt that we were there on some type of academic affirmative-action program. This is in 1976. I remember my first encounter with my first physics professor, who was a Nobel laureate. That was exactly what he said to three of us who had come to his office. He just screamed at us that we were diluting the quality of education for the entire university, and a shouting match ensued—with a few things I wouldn't want to repeat in public. The kinds of values that were so paramount in my life, that I felt were a part of my faith, a part of my relationship with God—which is the most important thing in my life, has always been, will always be—didn't seem to have a place in this environment where engineering and math and science came together. From a leadership perspective I quickly found myself figuring out how to take a stand and share with others how to assert their values. Whatever the values we may have, the things that we hold important don't have to be left at the doorstep when we come to the university campus. By asserting our values as members of that community—as members who actually did belong in that community even though others told us we didn't—we felt that we could enhance that community.

So if Berkeley had been more hospitable to you as a woman and as a person of color, you might have been able to be a conformist. But when it slapped you in the face and said, "No, you don't belong here," that helped you to decide, "I'm not going to play the conformist game in my career."

That's right. And that has led to a whole series of events over the years of my career. I didn't realize this at the time, but when I look back now I can see the catalyst. It began when I became an officer and eventually the president of the Black Engineering and Science Students Association. While I actually didn't want to be a leader—I'm an introverted person; I really prefer just to be by myself and not interact with a whole lot of people—I felt compelled and called; it was an absolute necessity for me to step up and voice my opinion and encourage others to voice their opinions.

For you, then, there was from the beginning no question of "value-free science." From the start value-free science was out the door—if you ever believed it in the first place.

Yes, that's right. It was very clear to me that conforming wasn't even a possibility. But what concerned me was that people around me, who, I would have thought, felt like I did, in fact were saying, "Well, you know, we need to be quiet here, we need to just work on getting our degrees so we can live our lives later." I really felt that that was a tradeoff we didn't need to make and shouldn't make. I thought we actually could shape that environment to be a better place, for us, for the people coming behind us, and for the people who were there with us. I wanted all of us to benefit from the diversity that comes when the mix of values, perspectives, and backgrounds is really a mix, instead of a whole bunch of people who are trying to forget where they came from so that they can graduate.

Why wasn't conforming an option? What values did you hold that made this point so nonnegotiable for you?

From my point of view it's a spiritual thing. My relationship with God has been such an integral part of my life from a very early age, since about the age of four. To use an expression that some people use in the Catholic community, it's not about a relationship between me and any other person, whether that person is my priest or my archbishop—it's about a relationship between me and God. And if I'm clear on what God is asking me to do, and I'm staying on target (which I haven't always done), I can feel it. That affirmation is right there. When I get into a situation like that one at Berkeley, it's very clear to me what to do. It's almost as though it's not even me. I'm the instrument through which the Holy Spirit is acting. And if I do a good job at being that instrument, we're going to kick some behind!

When you went on to Intellicorp and Apple and US West, was your Berkeley experience repeated?

No, it was different at each step along the way. It isn't even so much the places where I worked as it was the responsibilities that I was called to. The way I can best describe it is that along the way—forty-two years now that I've been blessed to be on this planet—I have been being prepared for so many responsibilities. Each time when I was called to something that I thought was way over my head, I found instead that the skills of the previous two, three, four, ten years were exactly what I needed to somehow figure out how to do it. I'd figure that one out, and then there'd be a new calling, and I'd look back and think, Wow, it's a good thing that I learned those things because I sure need them now!

So you've had a clear sense of God's preparing you for each new and harder task by using the tasks that had come before.

Yes, absolutely.

How far was it possible for you to find an integration between your faith and your work as a computer professional?

Everything going on in my life always feels integrated. There isn't a shifting of gears from church to office to home. They are just intertwined. For example, I play bass guitar for gospel choirs at church. When people ask about diversity in the workplace, I often use this example: Let's say somebody's going to put together a band. You can't put together a band with five drummers. Somebody has to be playing the piano or some other instrument … and you've got to have a bass. So what you naturally want to do is bring together a set of people, each playing a different instrument, yet each contributing to a harmony that sounds good, feels good, and draws a good feeling from deep within people's souls. That's how a band comes together; and when they put in the practice time together, they can trust each other musically to get the groove and go with it. You have to log some serious time together in order to do that. Well, it's the same thing in the workplace when you bring together a diverse group of people. When I go to band rehearsal, I really feel like I'm working on the same team skills that I use in the office when we are recruiting and blending new people into our team.

Have St. Paul's metaphors in the Christian scriptures—of each member's having a place in the body of Christ—informed and directed your collaborative style in the workplace?

For me the metaphor of the body affirms that style. When I read the New Testament, I find that so many of the writings are about how to cultivate a community. Each person is called in a different way, and as various people respond to their calls, together we're able to do the things that we're called to do as a community. St. Paul's words ring true to me. Some of that comes from the way I grew up with my sisters: in our family we all had very different skills and interests, yet we knew from early on it's a good thing for each one to be different. That I'm very introverted and very much into my books, that one of my sisters is very gregarious and has tremendous social skills—in our family that was all good. So I helped her with her homework, and she made me learn how to talk to people. Now, years later in the workplace, people can see that I've learned to be comfortable with other people.

How is it possible that you could develop this collaborative work style in highly competitive environments like Apple and US West? Didn't you face fierce resistance to your ethics of collaboration? How were you able to modify and adapt your ethics as a project leader in these firms?

Both at Apple and at US West I faced lots of resistance. Apple's environment was very combative. The way one person described it to me shortly after I arrived was, "If you are not willing to defend your idea in a design meeting, then you must not care about it." My view was that we're not there to defend *against* each other; we're all on the same team. We should be concerned, maybe, about the real competition, which is kicking our butts, but not about fighting each other. When I first joined Apple, I managed the Newton team, which was almost entirely composed of people who had been on the Mac team. It was a very small group at that point—eleven or twelve people—and almost all of them were doing things the Apple way. Now that I look back at it, it was amazing that Apple actually selected me, from outside the company, to be a manager of a team. But they were stuck—stuck in a spiral where they were inventing and inventing and inventing so much new stuff, but they were never going to ship anything. They wanted somebody to break that cycle.

I realized when I got there that one of the reasons the team only hired people they already knew was because they didn't know how to interview people. Their usual calibration was, "What have you worked on that I've worked on?" or "Who do you know that I know?" Since in my interview there were no connections in that space, they had no questions; they started asking me if I like to play basketball. So after I joined, I trained the team on how to do interviews. I said, "Interviewing is a skill we're all going to need to have, because we have to hire another fifty people here real quick. Thus the twelve of us need to get together and get really good at doing this." People could immediately see that interviewing well would be a good thing; it's an added value. After that I didn't need to convince the team members that we needed a new approach. Hence, while the rest of Apple continued to operate the way it did, and later the rest of US West continued to operate the way it did, in each case we started to cultivate a culture that was completely different: a team that was collaborative, a team that had a lot of trust, a team where people could talk openly.

One can't help but notice how similar your team-building approach to management is to basic Christian values of community. Is there an influence?

Oh, yes, absolutely. All the Christian values that the Lord has cultivated in me over the years—they come through every day, wherever I'm working and whatever project I'm managing. Likewise, the things that I learn from people at the office—the real life experiences of applying principles of love and community and fellowship—carry back with me to my church community.

Let's move on to Freshwater Software. There you have the chance to found and to mold an entire organization so that its mission reflects the values that you hold.

The focus at Freshwater is customer service; embracing that as our focus means that we do things differently from the norm. When we say that we focus on customer service, we really, really mean it. When you get everybody focused on that, a whole lot of other stuff just disappears. The territorial stuff, the departmental stuff, the turf stuff, the ego stuff—it just starts to disappear, if it ever started to appear. It doesn't have a place in our culture, because everybody's talking about the one important thing: customer service.

Can you describe an example of a principle from your faith that you have applied to management?

Resourcefulness is something I talk about all the time. I would love at some point to gather all the fantastic examples of lateral thinking in the Bible. One of my favorites is about a man who was paralyzed, whose friends wanted to bring him to Jesus. Jesus was in a house full of people who wanted to hear what he had to say, and all around the house was a crowd of people. The man's friends were carrying him on a stretcher, but they couldn't get into the house. So they went up onto the roof of the house, opened up a hole in the roof, and lowered their friend down through the hole. Jesus granted their request because of their faith—but also, I believe, because of the resourcefulness that said, "Find a way." Find a way. When you are called to do something important, don't let any obstacle hold you back. If you have to climb up onto the roof of a house and tear open a hole in the roof and lower your friend down in order to reach your goal, that's how you have to do it.

At our team-building meetings, we work on lateral problem solving. Practicing the exercises means that when the situations come up day to day, you're ready. Don't even assume that it's going to be easy. We're running an Internet software company. When we began, nothing we were attempting had been done before. The solutions of how to serve customers well in this context can't be found in any book. You have to come up with new solutions; you have to be ready to climb up on the roof and tear open a hole and get through. Those barriers can seem impossible sometimes, yet if you look closely, there's a way.

Some might say there's a conflict between too much attention to customer service, or too much money invested in service in general, and making a profit. How do you respond?

I think that's a way of thinking that has been cultivated to hold people back. The best way for me to answer is as I did at UC—Berkeley, when people said, "You can't succeed because you don't have the capability—because you're black or because you're a woman." I've heard such dismissive language before, and the best way I've found to respond is by delivering. When you deliver, a lot of the objections fall by the wayside. The resistance is not rational, it's emotional, so if I try to respond with a rational argument, it's impossible. And so what I respond with instead is results.

At Freshwater, for example, we've been profitable since October 1997. There aren't very many Internet companies that have been profitable. The fact is that we get over 70 percent of our new customers as referrals from existing customers. You can't buy that; no marketing dollars will give it to you. When people ask me about values and competition, I say, "Here's our experiment in running a business with a values-based culture that is focused on customer service, and these are the results."

Let's step back a little bit now from your career and talk more generally about science, ethics, and religion. What's your overall response to this issue: Does science stand in a fundamental tension with Christian belief and practice, is it compatible with belief, or does it actually support belief?

I'd say it is somewhere between compatible and supportive. The place where the answer is determined, in my opinion, is in the individual. If an individual is cultivating and treasuring a relationship with God, and that relationship is growing, deepening, and broadening day by day, such that his or her life is moving forward, then lots of things will be harmonious with it. What we learn and experience through science, I think, can contribute a whole lot to that harmony. So for me science is not only compatible but also supportive to my faith, since when my life is integrated, all these things fuel each other. When that happens, I continually have moments of "aha!"; these are rooted in connections between all the parts—between my spiritual life and business and technology and science and engineering.

I recognize that for other people the relationship sometimes seems more like friction. Religion represents for them a data point that is counter to their other data points, and now they have to resolve the tension. But I think of it much more as an individual choice that people make about how they integrate new stuff into their lives. The task of integration does draw me into questioning and querying and exploring and investigating more deeply. But the process feels good, and I'm actually happy when I'm working on it. It's like sitting down at the dining room table with all the pieces of a jigsaw puzzle I just poured out of the box. When I see all those pieces, I know it's going to be an evening of fun putting it together. But I can imagine that another person would look at all these pieces and say, "I have a thousand problems here to solve."

Someone asked me at the Aspen Institute a few weeks ago, "How much of something does a person need to be happy?" My answer to that was, "I don't need anything. The only thing that I could ever need I already have: it's inside me, and it's a relationship with God." Everything else is optional, because that's not only the most important thing, it's the only thing that matters. At some point early in my life I embraced it and said, "This is the way I'm living my life, this is working for me."

As you think back over your life and the lives of religious scientists you know, have you seen any signs of the scientific mind-set's being in tension with the religious mind-set?

No, no. I think that everything around us is around us as a gift, and we're invited to explore it. Suppose someone tells her children, "This backyard is for you to enjoy; it's a safe place for you to play games and pretend and run around." That parent's view is that the backyard is a gift, an environment created for her children, and everything there is to allow them to explore and create and invent and play and enjoy. If I scale that up, my view of God's gift to us is that so many of the things God puts around us and invites us to encounter are there as gifts. I've thought about this because I'm a single woman. I often travel alone, all over the world, but I never feel afraid. At certain moments, when I'm walking down a street and I realize I'm in more shadow than light, there's an awareness—but it's not fearfulness. When I go to Tanzania or Kenya, by myself for the most part, I feel comfortable because it's God's world. It's not that bad things don't happen. I know that people get attacked, raped, killed, or beaten up. All kinds of things can happen, but I focus on the fact that God continually invites me into exploration and adventure and education.

You know that God's going to take care of you.

Yes. And if I use common sense, which God also gave me, then it'll all be good.

Do you think that the way we learn things in the religious life is very different from the way we learn them in, say, the field of engineering or software design? Is it a different cognitive mode, as some would say?

These various modes may be different, but they are not incompatible. Think of communication. If you're having a conversation with someone in person, you're communicating verbally and also with body language, with inflection, and with vocal quality. You communicate with everything available.

When we were working on the Newton—to take another example—trying to create the first personal digital assistant, the main challenge was, how will people actually use it? We could put twenty more features into it, but if the users aren't going to figure out the first ten features, we don't need the additional ten. And if its usability isn't going to compel them to engage it as a tool, then they're not even going to turn it on. User interface and design and architecture all have to come together into a single combination.

In each case all these various modes are related to each other: the analytical approach, the intuitive approach—they're all pieces of what we're able to use to do the stuff that we're called to do.

I'd like to pursue a few questions about the field of human-computer interaction and the ethical issues it raises. Let's start with the generic issue: our generation and the generations coming after us are more deeply intertwined with computing devices than anything that we could have conceived even twenty years ago. Do you think that this fact has advanced, detracted from, or simply been neutral for the quality of human existence?

I think that advances in technology are very positive. In a lot of ways it levels the playing field, within the United States but also globally. The Internet, as a platform, is now available to anyone in the world who can get the technology that allows access. This means that, for example, where I'm working in Tanzania—in this rural village where there's no electricity and no telephones—they now use e-mail. They don't need to wait for the phone infrastructure, which (it seems) is never going to get there. On the other hand, e-mail via satellite and the Internet is amazingly inexpensive.

When we set up that first pilot Internet station at a village school in Tanzania, I didn't count on what would happen. The parents had been wrestling with the question "If we want a good education for our children, do we have to send them to the city, away from their families?" Now that the computer in their rural school has e-mail and can access the Internet, they're recognizing that their children have an opportunity for a better education than they had thought possible before, so that they don't *have* to send them away. I hadn't thought about it that way at all—I just showed up with a laptop and an antenna and started to hook them up. But when you put the tools into the hands of people, they figure out how to use those tools. They will come up with ideas that we would never have thought of.

So the spread of computing technologies can have very positive social impacts. Some claim that the expansion of technology will help transform unjust political structures around the world and will open up new democratic possibilities. Does that claim go too far?

That's a little further than I would take it—but not necessarily because that's the wrong direction to go. What I can say is that computing technology is a tool that is universal, like a knife. The Swiss Army knife that I own has been used for very different tasks than the Swiss Army knife that I gave to a friend in Tanzania. I know that we will use them in different ways, just as I know that if I gave a Swiss Army knife to an angry young gang member, he might use it in a different, maybe very negative, way. But it's still a knife.

I think the Internet is a kind of universal tool. Everyone knows that on the Internet you can distribute pornography or you can distribute access to encyclopedic educational information. Both are right there. You can contribute or you can receive. What a person's values lead her to do, and what her relationship with God compels her to do, she'll do with that tool, just as with the knife.

Within the United States do you think that the mediation of so many of our communications through computers has any negative impacts? Should these impacts be a cause for concern as computing becomes more ubiquitous?

When I was a kid growing up in Berkeley, my mother would take us to Louisiana every other summer to spend some time with our family. My grandfather lived in a small town, and they had a party-line telephone connection. It was a rotary dial telephone—really big and heavy—and you only needed to dial four digits to reach anybody in that county because there were so few phones. I noticed that when people would talk on the phone, they would talk loudly, in short staccato sentences. I thought, This is the same kind of phone that we have in Berkeley, but people here don't feel as comfortable with it.

Over the following years, that changed. Everybody back there now uses the phone in a much more natural, comfortable way. So my view of ubiquitous computing and e-mail and chat rooms is that people are still getting the feel of it. We're not there yet as an entire world. When I get e-mail from my friends in Tanzania, it has the flavor of my grandfather's early phone conversations; it's as though we can't have any extra words in the message, because this is costing a lot. Computing isn't there yet, but in time it will be everywhere, and ubiquitous computing devices will add value.

Now a final question on ethics and the future of computing. Researchers in artificial intelligence and robotics foresee building voice-activated, human-looking robots that may well become a common part of human interaction within a few decades. That's going to raise the question of how we treat these robots. Many suggest that at some point we will no longer be able to draw such a sharp distinction between robots and humans, and some say that at that point we should begin to speak of "them" as people like ourselves. Do you anticipate that sort of outcome, or do you have some reason to think that the qualitative difference between humans and computers will always remain?

I definitely do not have a final answer. I can speculate a bit, reflecting on the idea that God created all the living things on this planet. God created the plants, and I wonder sometimes about how we treat plants. God created all the animals, and I wonder sometimes about how we treat animals. God created all the people, and I often wonder about how we treat each other as people. While I don't think that we can be Creators, I do think the things that computer engineers invent, using the talents God has blessed us with, are often the things that we are called to contribute as the community of people worldwide moves forward. But, in the same way that I wonder about how we treat living plants and animals and people, I wonder about how we are being as stewards. How will we take care of the things that the Lord has given us these skills to invent? I do think that there's an important distinction between being a Creator and being a person who can build things. I don't think we are trying to make ourselves become God; instead, we are trying to aspire to do what God calls us to do.

It's a question of stewardship, because each thing has been entrusted to our care for some reason. We may not even fully understand what the reason is, but so many things in my life have shown me that God is not just playing around. There's a serious plan going on here; it's got complexity far beyond what I can put in my hand. But stewardship is not a small thing, and I think that sometimes we forget that.

5. Ursula Goodenough

Ursula Goodenough is an internationally known cell biologist. Born in New York City in 1943, she is a professor of biology at Washington University in St. Louis. She was educated at Radcliffe and Barnard (B.A., zoology, 1963), Columbia University (M.A., zoology, 1965), and Harvard University (Ph.D., biology, 1969), did two years of postdoctoral research at Harvard, and was assistant and associate professor of biology at Harvard from 1971 to 1978. Goodenough's research has focused on the cell biology and (molecular) genetics of the sexual phase of the life cycle of the unicellular eukaryotic green alga *Chlamydomonas reinhardtii* and, more recently, on the evolution of the genes governing mating-related traits. She has served

---

Philip Clayton interviewed Ursula Goodenough; Helen Bishop and Zach Simpson edited the interview for publication.

in numerous capacities in national biomedical arenas, and her widely adopted textbook, *Genetics,* is considered a classic in her field.

Goodenough joined the Institute on Religion in an Age of Science (IRAS) in 1989 and has served continuously on its governing council and for four years as its president. She has chaired three IRAS annual conferences at the Star Island Religious and Educational Conference Center in New Hampshire and serves on the editorial board of *Zygon: Journal of Religion and Science.* Goodenough is in international demand as a speaker on science and religion issues. Her best-selling book, *The Sacred Depths of Nature,* has led to extensive speaking opportunities in the United States, on television, and in interviews on National Public Radio.

She is an intense and restless discussion partner. Her sharp intelligence probes the questioner as thoroughly as it probes the fields in which she specializes. She is also a passionate biologist: no one can walk away from a discussion with her without realizing that unicellular organisms are fascinating evolutionary products. Always in conversations one senses that she brings an almost spiritual intensity and interest to the study of biological phenomena and an earthy, biological focus to discussions of religious topics—precisely the balance that is represented in her actual position on these topics.

Maybe the best way to begin would be to have you briefly describe the evolution of your research interests from the graduate level to today. These interests have led to your being recognized as one of the leading contemporary cell biologists. It might be interesting for people to be able to read about the evolution of your scientific interests.

When I was a graduate student in the sixties, I worked in two laboratories: one with Keith Porter, who was one of the eminent electron microscopists of his day, and one with Paul Levine, who was an eminent researcher doing studies of photosynthesis on a green alga called

Chlamydomonas. *Chlamydomonas* has an excellent genetic system, and in the days before we could do molecular manipulations, electron microscopy was one of the few ways you could figure out which genes were involved in a particular biological process. Most of my work focused on correlating the structure of the chloroplast in this algae, as seen by the electron microscope, with the biophysical properties of various photosynthetic mutants, such as those that lacked particular cytochromes, particular aspects of the light-harvesting apparatus, and so on. The technique was a real gold mine, and I made a number of discoveries using it, as well as receiving a wonderful training in the interpretation of ultrastructure.

What turned your attention in the direction of cell biology in the first place? Did you have classic scientific training?

I took my first science course when I was a sophomore in college. Before that I was trained only in literature and social sciences. Taking my required Biology 1 course, I was completely stunned by all this new information; when I was eighteen years old, I literally had no idea that things were made of atoms. I really liked getting down to that deeper level of scientific understanding and found myself particularly fascinated by genetics and cell biology. All of us in science tend to occupy that particular level on the reductionism-holism scale where we feel most at home and most able to make a contribution. Cell biology was a level that was just coming in at that time, since the genetic code had only recently been discovered. It was one of the frontiers of scientific thinking at the time, and it was exciting to be situated on the cusp of knowledge and innovation.

But why this particular frontier?

It really seems a matter of natural inclination and ability. Some people are really drawn to biochemistry and reaction kinetics, others to ecosystems, and my interests fell between those two poles. I also found I had a particular gift for being able to think about biology at the cellular level. I could look through the microscope at *Chlamydomonas* swimming around, or look at pictures in the electron microscope, and a number of things would come together for me. Along with the insights came ideas for experiments that would test the insights.

I think outsiders to science aren't aware that particular scientists have natural gifts for perceiving and thinking about the world at some particular level. Perhaps it's akin to having an ear for music.

You certainly figure this out once you've been around great scientists a bit—you realize that they're just really good at what they're thinking about. There is definitely a sense in which observation and discovery in science is much like having a musical ear. You simply begin to "hear" things that are not apparent to others who are observing the same phenomenon. Of course, there's some circularity here: if you've thought about an area for a long time, you'll naturally get better at it. Over time you come to know, in the whole array of things that are going on in a complicated system, what is worth pursuing and how to study it in a way that yields greater understanding. I think much of good science hinges on this natural ability to listen to patterns and rhythms in whatever is being observed.

There is something particularly intriguing about the cell; it's the first unit at which the complex processes of biology are visible. We know these processes at the level of bodies and ecosystems, but there's something particularly fascinating about being able to trace the interactions at the most fundamental level.

Absolutely. Although *Chlamydomonas* is a single cell, it's also an organism—or a dynamic representation of one; everything this particular cell does is also what organisms do. Since I've specialized in its sexual behavior, you might say I'm really an ethologist: I study an organism in its natural environment; it's just that I'm working at the smallest level at which you can study an organism. Single-celled organisms have all the properties of being a species, including their sexuality, their origins, living within a particular habitat, and their highly adaptive way of performing tasks. Formally, it's exactly the same thing as a mouse or an eagle.

Can the study of simple organisms shed new light on what are the minimal conditions for being an organism? What are the similarities and what are some of the differences from a more complex organism such as a chimpanzee?

Well, the key difference is that all the cells in the chimp are tethered by the overall strategy of the organism, so the trillions of cells become bit players with their own particular tasks. As [the zoologist and author] Richard Dawkins would say, they all have the same ultimate goal, to get the organism to work, whereas in a single-celled creature the cell and the organism are one and the same thing. In this case you can more clearly see the individual functions of organelles and how each contributes to the organism's adaptive response. In multicellular organisms the individual effects that are so evident at the single-celled level are blurred.

The multicellular strategy has created organisms that are particularly interesting to us because we're one of them. But it's not necessarily the most dominant strategy in terms of proliferation or survival. Prokaryotic bacteria also do everything as single cells, and there are more bacteria in a spadeful of soil than there are humans on the planet, so the single-cell concept is by far the most successful and adaptive idea that has come through evolution. Looking at the success of single-celled organisms calls to attention how derivative cell functional specialization is, even though it may have created organisms of particular interest to us.

Can you summarize the central focus of your own research?

Throughout all my graduate training, I found myself particularly interested in the mating reaction of the *Chlamydomonas*. There are two mating types of these eukaryotes; when you mix them together, they undergo a very elaborate behavior: sticking to one another, fusing, forming a diploid cell, forming a thick cell wall around themselves, and becoming a zygote, until finally that zygote undergoes meiosis. The organism essentially replicates much of eukaryotic evolution in its mating sequence. Beginning with my first appointment at Harvard, I found myself captivated by this reproductive dance and sequence, which is instructive on so many levels. As the field has evolved, I've been able to apply a series of new technologies and to obtain deeper and deeper answers. For example, using contemporary genetic analyses we have now discovered the specific genetic disruptions within a series of mutants that could not reproduce, which I isolated many years ago. Sometimes one has a hunch about how things work but no way to assess that hunch experimentally until much later.

Can you talk about the role of these hunches: Are they more like educated guesses or like intuitions? Science textbooks rarely talk about hunches, and many nonscientists believe they play only a minimal role in a successful research program such as yours. If hunches are important, is there any connection with the sort of intuition that religious people talk about?

Of course, there are the rather quaint examples of hunches, like Isaac Newton's apple. But there are also many true stories about particular insights that come to scientists at a particular time, which then led to important new tests of the insight. Such hunches are not rare ... but, unfortunately, they are not often correct. I had a hunch this morning and then realized that there was a tube in my rack that would tell me whether I was right or not; I then looked in that tube and found I was completely wrong. In contrast to theology the satisfaction comes from a scientist's ability to confirm or falsify such hunches. And it's these hunches, micro or macro, that often form the foundation of good science. Thus the hunches are rather different in science than in theology: in science the data constrain our conclusions much more strongly, so that the leaps are much less spectacular.

Some theorists of science, such as Karl Popper, claim that the initial hunch might be similar for a theologian or a scientist; the difference is that, when it's science, you can check the tube. Whenever you can't check a hunch, even in principle, it isn't science.

I agree. At that point, to the extent that the hunch continues to be something that works for you, it becomes a belief. Belief is the resting place of untestable hunches.

And then, over time, can that belief become more and more justified, so that eventually it may become part of a really fundamental theory in some area?

Perhaps in theology. If you can't test the hunch in theology (which you can't), what you can do is embed it in other theological premises, or in the core grounding of the tradition, until the hunch becomes a candidate for belief. But the process by which Mother Nature says, "Sorry," as it occurs in the sciences, is simply not available to you. Unfortunately, in theology she can't also say, "You are exactly right!" By contrast, in science the voice of Mother Nature is always there, constantly speaking to us in the form of observations. That's the part that people don't understand about science when they complain that "scientists keep changing their minds all the time" and conclude that the scientific enterprise is just a house of cards. But that's not accurate: the observations may change as our ability to observe improves, but the essential notion of that voice of confirmation is always there.

In your book The Sacred Depths of Nature, *you speak of those moments of discovery or confirmation. You suggest that they can be akin to moments of great religious experience or insight. You even use the same phrase to describe observing a signal transduction cascade and looking at Aztec ruins by moonlight; as you put it, "Same rush, same rapture." Can you describe what such a moment is like? It must be an amazing experience.*

It's definitely a fascinating moment. But with maturity you come to realize that the greater miracle—if you want to use that word—is that Mother Nature does anything at all, and that, just for a moment, you're privy to certain aspects of Mother Nature herself. It's amazing that we're clever enough to examine her and to come to know some of her aspects. That with the help of the instruments that we have come up with, and building on all the progress that earlier scientists have made, we're able to make a mental inference and then to think of experiments to confirm it—that's a bit of a rush, yes. The real rush, though, is what's still out there, potentially eluding our best attempts at observation so far. The sense of awe does not just come from what we as scientists find; it comes from the fact that we find anything at all.

Are there gender differences among scientists concerning how important they think these "hunches" are to the doing of science? Some people have claimed that there are some important differences in how men and women approach scientific work.

I'm not entirely persuaded by that idea. The first piece I ever wrote for *Zygon* was an article entitled "Creativity in Science." As background research I interviewed a large number of scientists about their "eureka!" moments, as well as drawing on famous descriptions of such moments from the literature, and I found no significant difference between men and women. In recent years the most famous claim that such differences exist is Evelyn Fox Keller's concept of Barbara McClintock's having a "feeling for the organism." But limiting this idea to women is off target: I know plenty of men who have a terrific feeling for the organism and a lot of women who don't have it at all. In the end, having or not having this experience has much more to do with one's temperament or level of empathy, and with one's talent, than with one's gender.

You have become famous as a leading scientist who is also willing to discuss religious topics, and in recent years you have made significant contributions to the international dialogue between science and religion. Do your ongoing interests in spiritual topics make you a theologian of sorts?

My initial tendency is to be hesitant about using that term. But I have become much more comfortable with thinking about theology after a conversation with Karl Peters, who was one of my early mentors. Karl referred to himself as a theologian, and I said, "Karl, that's really interesting that you call yourself that because I thought that a theologian was a theist." He replied, "Well, the etymology of *theos* means 'that which is ultimate.'" Consequently, I think of theology as "the philosophy of ultimacy" and hence as a method for asking questions about ultimacy and reflecting on how one might address such questions. From this perspective I might feel comfortable calling myself a theologian. Ultimately, what I'm doing isn't theistic, but it is, in this sense, *theological:* I'm attempting to talk about questions of ultimacy.

I take it that you intend by the word theologian something rather different than the term has traditionally meant?

Yes, the differences are important. Whether because of my temperament or because of my training as a scientist, I'm perfectly content that others can speak more comfortably about what ultimacy is or might consist of. I don't feel comfortable making such strong assertions. For this reason I began to speak of the notion of a "covenant with mystery." Some people have complained that this response is a cop-out, but I introduced the term as a sincere attempt at honesty. What came before the Big Bang? Why is there anything at all? These are questions to which I cannot give a fully satisfactory answer. It's from this sense of mystery, of *not* knowing, that I derive the covenant. Rather than being disturbed by a lack of answers, I'm enchanted by it. My own religious life, then, and my life in general, move along in a way that is very fulfilling without having a developed theology. And yet, unlike the professional atheist, I don't regard those who *have* a specific theology as involved in some sort of absurd activity or search. It just isn't a search that I'm engaged in. As one who is in covenant with mystery, I am a bit at odds with full-fledged theologies.

How can we better understand the contrast that you're describing? Perhaps there are people whose sense of the sacred involves being drawn to specific beliefs, recognizing that they hold these beliefs, reflecting on them, formulating them, raising philosophical questions about them, and writing theologies about them. And then there are people like you for whom the sense of the sacred—whatever that term finally means—has less to do with the cognitive, belief-forming side of themselves and more to do with the nonverbal, affective, experiential, or holistic side?

Yes, that's a helpful way of putting the contrast. I would like to add something significant to what you call the experiential or holistic approach to the sacred. What is sacred for me is *what is*; I don't then have a religious need to move on to "the why," to look for ultimate explanations for causality or purpose, or to invoke an ultimate external cause. There's so much that I find that is sacred and deeply mysterious, even apart from such analyses. In this sense I truly feel that my cup runs over, without my having to be concerned about theological issues in anything like the traditional sense.

When you say that you don't need theories about ultimate causes or about divine purposes, are you also saying that you don't need to appeal to the notion of transcendence when you think about what is sacred?

Transcendence is a very interesting concept. I've written articles both on the idea of transcendence and on the notion of causality. The latter is actually a sort of dyad, consisting of both causality and subjectivity. Think, for example, of the inevitable human search for causes. In fact, all organisms seek causes; it's an instinctive process. Where we humans are unique, however, is that, when we can't ascertain a cause, we imagine one. I use *imagine* not in the pejorative sense but in the sense of imaging: it's part of a hunch—or, perhaps, part a series of traditional or inherited hunches. What is interesting is to view the theist, nontheist, agnostic, and atheist in light of this instinct. For theists, if life and existence are to be meaningful, it's crucial that they have some answer to questions of ultimacy; these answers are essential for their life to feel meaningful to them. The atheist has an entirely different belief system, but it's still an answer to an ultimate question. I propose that there's a spectrum of human natures: some people really are existentially miserable without some ultimate causal explanation, and other people don't seem to need one. I seem to be in the latter group. In my view, what is important is that we all get on the same page about *what is*. Our ultimate causal explanations for what is can be much more personal and diverse, with the diversity of theological explanation in fact being something that we should treasure.

Would you say something similar about the transcendent? Perhaps for some people there wouldn't be a sense of meaningfulness unless there is transcendent meaning, whereas for others no answer needs to be given in terms of a transcendent source?

The best place to start is, likely, to inquire as to what actually constitutes the transcendent. I entitled my essay on this topic "Horizontal and Vertical Transcendence," which gives some indication as to the distinction I make within the concept of transcendence itself. The distinction is gleaned from a professor at the University of Washington in Tacoma, Michael Kalton, although I develop the argument in a somewhat different direction. I argue that vertical transcendence—the experience of going beyond specific experiences, the experience of discovering overarching meanings—is accessible to most, if not all, of us in the form of art. Some people continue on in the process of vertical transcendence, using the categories of beauty and coherence, until they arrive at more theistic concepts and experiences. But all people align themselves to some degree along this vertical axis, which is typified most brilliantly by art. It's just that, for some of us, God-talk is not how we describe our experiences, whereas for others it is.

I believe the experience of vertical transcendence is somewhat universal. One can then ask whether there is such a thing as horizontal transcendence at all. Is transcendence only about a vertical, hierarchically ordered reality or can we also have transcendent experiences just by taking in things the way they are, without imposing on them levels or degrees of reality? Can we ultimately appreciate reality by *not* going beyond it? That's the question of horizontal transcendence. Horizontal transcendence involves the appreciation of *what is*, of what we already know and see.

Is art always vertical? Couldn't the artistic experience, or indeed the entire aesthetic dimension, be part of horizontal transcendence?

I interpret most art as expressing vertical transcendence, and I view much of our experience of nature as emphasizing vertical transcendence as well, because these sorts of experiences emphasize a dimension beyond the real, the ordinary. In these contexts the lofty mountain peak, the perfect flower, and the ocean are taken to pointing beyond themselves to something else, something like a Platonic form. If our only experience of nature is aesthetic, however, we're really in deep trouble, because then we're in danger of only working to preserve those parts of nature that are aesthetically meaningful to us. In fact, most of nature is messy; most of nature is swamps and rather unpleasant creatures bubbling around. Most manifestations of nature are not particularly inspiring. We can't always be looking *beyond* nature to value nature itself. Nor can we only value nature because *we* value it. Which is why, I believe, we should defer instead to a child's experience of nature. Children aren't found standing on the ledges of cliffs contemplating distant vistas; you find them continually rummaging around in the mud and just being creatures. They are our best guides to horizontal transcendence. They don't "appreciate" nature in the aesthetic sense, simply because they're too busy actually appreciating it: really relating to nature, discovering it, responding to it. It is my sincere hope that we can cling to that childish incipient relation to nature that we all once had and that we can rediscover transcendence in this sense. When you see kids out in the woods, they're having transcendent experiences all over the place—horizontal ones. They are not attempting to look beyond or behind nature; they're *with* it and living in full relation to it.

That's why you emphasize the experience of wonder, is it not? because wonder represents the kind of transcendence that is at our fingertips; it's horizontal, here and now, right in front of us. Wonder plays a large role in your understanding of religion.

Exactly. Of course, wonder can accompany both horizontal and vertical transcendence: you can certainly be overtaken by wonder while hearing a Gregorian chant or a string quartet or viewing a piece of art. Thus you could say that I'm aiming to expand the opportunities for humans today to have religious experiences. Many people are hesitant to use the word *religion* because of its general connotations of piety. But when we do this we miss out on a great deal that's important. Thus I hope to expand what is generally meant by "religious experience," so that the term can also include the wonder that belongs to horizontal transcendence. The result is a notion of religious experience that's more inclusive of our simple relationship with nature, allowing us simply to see it. This type of relationship can be just as spiritual or powerful as those experiences that take you beyond nature itself.

How has your understanding of spirituality been affected by your work as a cell biologist?

Almost all my work builds on the biological understanding of how life works and how it evolves; indeed, those are two sides of the same coin. It is the remarkable beauty of the cell, of everything that is, coupled with the improbability that life would have originated in the first place, that continues to draw me to spiritual issues. Additionally, the fragility of existence, its deep interconnectedness and interrelatedness, and the amazing genetic kinship among living things—all of these form the basis for my constant sensation of wonder and awe. Indeed, understanding doesn't seem to diminish my sense of wonder; it only enhances it. And this feeling of wonderment is the foundation of my nontheology, as it were, my religious quest.

That's beautiful. Since your last answer begins to point toward it, can you talk more about the ethic, the sense of value, that you accept? There is a proposal for how we should live that seems to be basic to your spiritual vision.

There are at least two broad categories of ethics: one is how humans should behave with respect to one another, and the second is how they should behave with respect to the earth. For me the most important insights have come from the understanding that we are evolved apes. We have psychological characteristics that are emergent, but they emerge from systems that are deeply homologous to our own. Despite the fact that we have these emergent sensibilities—ones that we are justifiably very interested in—a lot of human psychology is informed by, and constrained by, our evolutionary history. Many issues that we grapple with morally have major antecedents in our evolutionary history: violence, infidelity, suffering. For example, we all come into the world with violent tendencies, despite the fact that our psychology is such that we certainly don't want violence to happen to us (although it might be permissible were it to happen to someone else). Consequently, evolutionary psychology can explain much of why we do what we do and why we are the way we are. Evolutionary explanations can also be misapplied, of course, as when the rapist goes in to the judge and says, "My genes made me do it." That, as far as I'm concerned, is misguided, since our knowledge can trump our genes. Still, one's view of human behavior is transformed in important ways when one starts with the understanding that many behavioral tendencies come to us via evolution, rather than via the Fall or a defect or as an aberration. In the instance of the Columbine massacre, for example, if those children had been living in a culture where alienation and violence were understood as natural tendencies to be controlled, and not as an aberration or a perversion—or, worse, romanticized—then I believe that such atrocities may not have happened. The ultimate "oughts" may come from our psychology and culture, and they may be arbitrary from a scientific point of view. But they are not divorced from the "is," from who we are, and this "is" must also be taken into account.

You describe a sharply realistic ethics, yet it's not one that just accepts the status quo or claims that all our innate tendencies are fine.

It seems to me that we are at a very interesting moment ethically. In part we realize that we might be apes after all, and that raises a number of moral quandaries, some of which border on the absurd. At the same time the old criteria for arriving at moral judgments, that is, as written on tablets of stone, are being eroded, and we're beginning to understand that our emotional systems are not designed to always act in the best interests of others. With the loss of both the old codes of conduct and the faith in the goodness of human nature, it seems as if the whole planet is wandering around like a boat without a keel. There are just a lot of oars going and a lot of spinning boats. If we could really understand who we are and how we got here, that would have a major impact not only on our moral codes but also (and more important) on implementing them. Science can be helpful here; it can help to give an account of our history, of why we are the way we are.

And is that how you make the transition from human relations to relations between humans and other animals?

Yes. The next issue is to ask what we are doing here and how we are to position ourselves with respect to other creatures and the planet. Here again the scientific understanding has a great deal to tell us about what really happened, and how miniscule we are, in light of our broader evolutionary history and in comparison with other organisms on the planet. That understanding is in the same category as understanding that we have mammalian, and even reptilian, emotional systems. That understanding in and of itself doesn't generate solutions, particularly at this point in our history. For instance, a friend calculated that if there were the decision to go back to being hunter-gatherers, 99.8 percent of humans on the planet would have to be killed, because a hunter-gatherer society on this planet could sustain only about ten million people. That understanding doesn't generate solutions, but it does challenge anthropocentricity. Anthropocentricity is the other end of scale from claiming that humans are "just great apes," and it is equally dangerous and maladaptive. There has to be something in the middle, and finding it is one of the great challenges of our time. We have to realize an ethic that both accepts our smallness and acknowledges our present situation in the world.

Before we close, I have to ask you one question that arises out of reading your work. Could one retain a sense of the sacred depths of nature if one became convinced that physicalism is true? What if I became convinced that all properties of nature, from single-celled to complex organisms, were ultimately explicable through a process of derivation or deduction from the basic physical constituents of matter? Would that leave untouched my sense of the sacred depths of nature or would it have a dampening effect?

We must first clarify what exactly *physicalism* means. I am a physicalist in all but one sense of the word, so clearly my book and my work presuppose at least a form of physicalism. However, despite my brain's being simply the composite of trillions of neurons, I do believe that there are new emergent sensibilities that have formed in humans that are not present in our antecedent evolutionary history. Since the sensibilities are dependent upon, and will ultimately be understood as manifestations of, the physical, I have no need to argue for a more dualist or supernatural picture of the mind. But I do robustly believe that the abilities of the mind are something new and different from all that has happened before in evolutionary history. They generate certain behaviors that are not present in our preceding evolutionary history, such as caring about other people and creatures. Without that ability to care, for example, you would never even be concerned about what was sacred and what wasn't. So it seems that the physicalist posture not only allows for a feeling of wonder but, in a certain sense, helps to explain how wonder is even possible in the first place.

Emergence is a fascinating concept. How fundamental is that term to the view of science and religion that you are suggesting?

A slogan that works enormously well for emergence is "something more from nothing but." For example, there are proteins called actin and myosin, and actin by itself does a lot of things in the cell without any myosin around. And myosin, by itself, does a lot of things without actin around. But when you get a creature that has genes for both actin and myosin, then these two proteins interact with one another in such a way that the myosin hydrolyzes ATP [adenosine triphosphate, an energy source for cell metabolism] and moves along actin fibers. What you then get is an emergent property called motility, even though the motility is nothing but actin and myosin and a simple energetic expense. This is essentially the foundation of muscle contraction. Rather than relying on an appeal to a supernatural cause, then, what occurs is the formation, the emergence, of a new property, motility, which is then subject to natural selection. Emergent properties are not just a theological concept, then. We use them in the lab to describe the organization of complexity on multiple levels. We use it to describe things that interact, such that the result is something more than its constituents. In these cases traits evolve that then become the focus of selective pressures. In this sense the brain itself is an emergent entity: it's the unique combination of a set of processes and physiologies that were not present in prior evolutionary history. Much of primate research has borne out the crucial role of such emergent structures and properties. As a result the whole notion of physicalism is not problematic for me. One can still have uniqueness and creativity emerging out of a fully physical substrate. Evolution depends fundamentally on the "something more" aspect of emergence.

Could it be that among the "something more" that emerges are those properties that we experience as psychological states, and those things that we speak of as values, and perhaps even the sacred itself?

Actually, I would turn it around the other way: having these specific psychologies makes it possible for us to hold such attitudes as gratitude and reverence and care and so on—states that you really don't see even in our closest cousins. The ability to nurture, for example, which is itself an emergent behavior, arises through evolution; its cognitive components stem from processes occurring in the brain. But we humans take nurturing far beyond child rearing. Nurturing, for example, becomes a key aspect of dealing with other people—even sometimes in politics. We've taken an essentially inherited attitude and expanded its original usage. In much the same way we've adapted the natural values of gratitude and reverence, and even the sense of awe, to the fully human phenomenon of spirituality. This adaptation helps to produce sense of the sacred.

Early in our discussion you turned the conversation back to the fundamental question of reality: What is it that ultimately exists? I want to ask in closing: Does reality include value? Is it true to say that "what is" is good?

For me, nature simply *is*. It is value neutral. It's not evil, it's not good; it just *is*.

Would you say the same thing about the predicate sacred? "What is" is neither sacred nor not-sacred; it just is. We may view it as sacred, just as in some cases we view it as good or bad.

There are two senses in which one might speak of "the sacred," and the answer is different in the two cases. In the primordial sense "the sacred" precedes any kind of judgment at all and has to do with what is; I'll come back to this sense in a moment. But then each of us, in attempting to locate or designate what is sacred, has to make distinctions and judgments as to what we take to be sacred. We can't simply designate everything as sacred. There are lots of things that we have to do as organisms to survive: we have to eat, we have to kill. Thus *sacred*, taken as a judgment, is a term that always demands qualification or choice. Calling something sacred involves choosing between competing demands for value and sustenance. Specific judgments about *sacred* and *profane* force us to divide up what ultimately *is*. In this sense *sacred* is ultimately something that we perceive with our psychology, much like the terms *good* and *evil*. But now let's go back to the first and more basic use of the term *sacred.* I take nature, understood as all that which *is* and *was,* to be especially sacred. In the final analysis this is a recognition that goes beyond simply attributing value to things, for it values that which precedes all things. Whatever religion or spirituality might mean in the ultimate sense has to do with this most fundamental level.

6. Thomas Odhiambo

It is with deep sadness that we observe the passing of the world-renowned entomologist Thomas Odhiambo, who died as this book was being prepared for publication. Dr. Thomas Odhiambo was a native of Kenya, a Christian, and a student of African native religions. He received his M.A. in natural science and Ph.D. in insect physiology at Cambridge University. He selected his field of research—entomology—in order to contribute to the development of increased food production and improved health in rural communities in Africa. In 1967 Odhiambo responded to the problems faced by African village farmers by founding what was to become the International Center of Insect Physiol-

---

W. Mark Richardson interviewed Thomas Odhiambo at the National Institute for Advanced Study in Bangalore, India; Kevin Lucid edited the interview for publication.

ogy and Ecology at the University of Nairobi. During his twenty-five-year tenure as the center's director, it grew into an independent research organization with a worldwide reputation. More than 150 African scientists were trained under Odhiambo's supervision with the goal of using scientific knowledge to support sustainable rural development and to alleviate disease.

Odhiambo later played a key role in the formation of the African Academy of Sciences and served as its founding president. In the last years of his life he participated in numerous UNESCO and Science and the Spiritual Quest events, including major international conferences at Harvard University, at UNESCO World Headquarters in Paris, and at the National Institute of Advanced Studies in Bangalore, India. The author of more than 130 papers and books, Odhiambo also wrote six children's books designed to educate, inspire, and entertain the children of Africa. He was a uniquely gifted speaker and a scholar whose integration of science and values exercised a powerful influence on students and colleagues around the world. In his January 2003 address to the National Institute of Advanced Studies, Odhiambo observed:

> Using the African holistic social framework as a reference starting point, it is becoming clear that a person is more than body. He is more profoundly complex than merely a physical entity—even one who possesses a full set of sense organs and a brain to manage and manipulate the avalanche of information being perceived by the body. The individual . . . has, in essence, a living-force which constitutes the transcendental part of the individual—comprising the mind, the intellect, and the soul. It is this transcendental component of a human being which confers to it the unique quality of human-ness—of *ubuntu,* of *dhano*. . . . When this *ubuntu,* this communion of interconnected minds, becomes inserted into

> the mind of the Supreme (God), there is no telling what heights of creativity, of discovery and invention, of innovation and improvisation, are likely to ensue!

During our many discussions Thomas Odhiambo spoke with that mixture of African and British accents that one associates with the speeches of Nelson Mandela. In fact, Odhiambo's demeanor, his distinguished gray hair, his poise and posture all created the impression of a man equal in stature to the great South African leader. In each sentence one could sense the mind and energy of a gifted diplomat, a man accustomed to conflict resolution and large public audiences, one who carefully weighed his words and never lost sight of the deeper vision that he carried. I will never forget the way he pronounced *ubuntu* and *Odhiambo,* with deep open *o*'s and *u*'s that will forever link these two words in my mind. Here was a man who had intuited the deeper unity underlying the various African religions, the unity of *ubuntu,* and who had found a way to harmonize science, religion, and values in light of this distinctively African insight.

As you know, the Science and the Spiritual Quest project has worked with scientists around the world, examining how they have been able to integrate the scientific method and mind-set with a larger spiritual picture of who they are. For many of these scientists this quest has meant reexamining the relationship of their scientific work to their moral and spiritual commitments. We are interested in hearing about how you have approached the task of integrating science and values, where you have been successful, and where you have faced challenges. Can you begin, though, by describing your background: Where did you study, and what were some of the important influences that led you to choose a scientific career?

During my early years I read a large number of books that were brought to me by one of my uncles. One of them happened to be a book by Alex Alexis Carrel, who was at that time at the Rockefeller Institute for Medical Research in New York. The book was entitled *Man the Unknown*. This idea immediately fascinated me—that there were so many things that were not known about man, biologically and in other ways. Much more important, I thought about how little is known about the relations between man and his god. This early reading opened up a whole new area of questions and interests for me, and it became apparent that I was going to be a scientist, rather than what I had thought I would become, which was a politician. These questions changed the whole perspective of my life.

A fortunate read at what sounds like the right time. Did your family immediately support this direction?

Yes, in a number of ways. I am very fortunate in having parents who gave me tremendous freedom. I was the firstborn in a huge family of eleven. In fact, I had to work, even though I was studying. From the time I was in upper primary school, I was working to support my own education and to give a little bit to support my brothers and sisters. But my parents were supportive in the sense that they gave me freedom to do what I needed to do, which was to pursue a top-rate scientific training. Notwithstanding this freedom, however, my father did want me to become an agriculturalist, a farmer. That was one of the things I did not want to do—even though I ended up eventually becoming the first founding dean of the Faculty of Agriculture at the University of Nairobi, in Nairobi. [*Laughs*] That was a part of the freedom too. Thankfully, I was able to mobilize sufficient intellectual, financial, and institutional resources to establish a world institute dealing with the role of insects in ecology. Since then, more than a 150 Ph.D.'s have been undertaken in this field in fifteen years. That has ensured that anywhere you go in Africa now, you will certainly find three or four doctoral practitioners in insect physiology or ecology or pest management.

Can you give an example of some events that stand out in your mind as particularly transformative for you—events that helped to form this deep concern that your work contribute to the welfare of others?

Two experiences from an early age come to mind. One dates to when I entered upper primary school. At that time one of my commitments was to be a catechism teacher in outlying schools. I used to walk about fifteen miles on Sundays to go and teach catechism, and back to school again. That process of religious teaching, which I did for about two years, has continued to have a deep impact on me.

The second one stems from the fact that, from the period when I was eleven years old, I did not only work for myself. In fact, from eleven years on, my parents didn't look after me in the sense of material things: I looked after myself, I paid my fees, I paid my pocket money. But I also gradually started paying for my other ten siblings. When I was in college, I was still providing for my family, so that my siblings sometimes regarded me more as a father.

These two experiences, I think, foreshadowed the idea that I have continually emphasized, that there is something meaningful beyond simply gratifying our own needs.

That principle is certainly apparent in your journey through life. It seems that you experienced something in your life in the family and community that played an important role in forming your core beliefs.

Society, for Africans, plays an extremely strong role. But there are now new challenges facing the Africa in which I lived my formative years. For example, until very recently we did not have the phenomenon of orphans. We never have had the problem of prisons. Prisons are a new invention in Africa. If there was an offender, the society found a way to deal with that person that brought him back into society; he was disciplined by society. But you were never thrown out of society. Often, if somebody died, or one or both of the parents died, the son or relatives would take care of the family. There would be almost no difference between the orphan and the child with living parents.

I recall your expressing this concern in the 1999 SSQ forum, when you presented your paper on the power of faith in science and spirituality.

Yes, very much so. At the SSQ India symposium, I also presented the results of a study of the very elderly in a district of Stockholm, Sweden, by Laura Fratiglioni of the Karolinska Institutet. The vital understanding we derive from this study is how crucially important are community connectedness and social interactions. What matters most is for humans to be embedded in social networks in a rich and deeply interactive manner. Social engagement that has this rich character mobilizes our cognitive capacities, leading to a sense of purpose and fulfillment, and a rich social environment is important for maintaining a proper psychological balance.

Have you personally experienced this phenomenon of community in Africa?

The experience is indigenous in African life. Social connectedness and integration characterize the context of life in Africa. I think the two are basic to the perception of Africa's deep and enlarged social space; they may indeed explain the astonishing resilience of African societies in the face of the horrendous pressures of the last millennium or so. Our experience of recurrent tragedy—the African slave trade, which ran for nearly seven hundred years; another three hundred years of colonial subjugation; long episodes of multiyear droughts; and the disease, epidemics, and famines over the last two centuries—has left roots deep within Africans as well.

Community connectedness, and the spiritual connectedness of the family and the community to God, runs through all humankind's relations, including its relations to nature. Relations in the family and relations to nature give added value and vitality to the individual in his quest for wisdom and knowledge.

There is something of Africa that stands out clearly in your description of the interconnection between life and God. Tell us more about these ideas and about how you have approached this theme in your career.

Before I went to the University of Cambridge for my university training, including doctoral training, I had already decided that the most important contribution I could make to Africa was to make insects the focus of my research. This ambition flowered during my Cambridge years; it became my dream to make this research an important focus for the whole of Africa. Why? Because insects are totally dominant in Africa. Our very major pests, both for crops and for livestock, are insects. The reason why many parts of the continent were never colonized was because of insect-borne diseases like malaria and sleeping sickness, or *Tripanosomiasis*.

In studying the smallest level of African life, you saw something much larger—a vision of what you hoped would lead to the resilience of African society.

I came to this discovery naturally. Historically, you see, most research in Africa's development had to be undertaken by army people: French army majors who left the war in 1944 and who joined the research groups dealing with the tsetse fly and *Tripanosomiasis*—sleeping sickness. They looked at how to control this menace in strategic military terms. Theirs was a "scorched earth" policy: clearing forests, moving cattle away, moving many people. That was their solution. It did not seem to me that this approach was sensible or rational.

Therefore, when I returned to east Africa, I immediately started a small laboratory research program dealing with tsetse, insect physiology, and ecology and behavior. That became a very major program, named the International Center of Insect Physiology and Ecology. Eventually, my researchers took it over. During some twenty years we developed systems for control of the tsetse fly and the disease it carries. The solutions we developed had almost no market, yet the system is so simple, so people oriented, that communities were able to immediately put it into operation and to clean up the countryside. That, to me, was the answer for Africa. You needed highly science-intensive methods for pest management but not necessarily highly technological solutions.

That distinction set you apart from your colleagues. How, specifically, did you approach this control system at the center?

We had discovered that certain African animals produce a chemical that is very toxic to the tsetse. And those are the animals that are kept in the general area where the tsetse will get their blood meal, even if they are not actually fed there. The first thing we did was to ensure that, when we are trapping the flies, we trap them along with those particular chemical elements, so that we could eventually synthesize the required components and put them into the environment.

We also rethought the trapping technology. When the flies fly into a trap, they can't get out again. Knowing that they will die when exposed to heat beyond a certain temperature, we designed traps that were white. What kills them, then, is solar energy. So they are eliminated without using any insecticide whatsoever. By doing this, and by letting the community itself devise those traps, which we also trained community members to design and make, we have been able to produce these traps at the cost of something like five dollars each.

The British government, when they came in, abolished all that. In fact, they also passed a law that anyone found doing intercropping would be jailed.

I recall that intercropping is a technique for which your center became very famous. Can you describe this new method that you developed? Also, what motivated the opposition by the government? Was their objection based on the sense that one had to conform to "proper methods"?

That's right. They had the idea that one must conform to standard commercial methods. The government idea was that they wanted to produce one crop on a plot of six acres. They did this by saturating the plot with commercial chemicals.

Still, certain investigators wanted to know what was the basis of the method that we had developed. We had discovered that these insects that became such a problem in farmlands did not overpopulate the wilderness. Why did the same problems not arise in the wild? The reason, we showed, is that in the wild, there are intercrops, that is, ecosystems defined by the interrelations between different species of plants and by their relations with the insects.

So we studied this interaction of insect and plant. We found, eventually, that insects are very much driven in terms of chemicals, rather than by vision. Their world is a chemical world. When we were able to work out the details of this chemical world, we immediately knew that we found the key. A certain mixture of chemicals, we found, would keep away insects. We have now gone a lot further with that, working with crops of maize, sorghum, and millet. If you take a field that is planted with maize or sorghum, and then introduce particular plants that in the past were regarded merely as weeds, you can totally keep insects away without using any chemicals at all.

The chemical structure was key. Today these ideas are just now making it into the fields and beginning to have their effects on the environment. We have been working for five years on this problem alone. More broadly, over the last ten years we have been researching the characteristics of particular crops at the Research and Development Forum for Science and Development in Nairobi and Kisumu.

In this work you have integrated a scientific approach with a highly politicized question. It's a powerful example of a natural, cost-effective solution that stems from intensive study in the sciences and that can be introduced into rural technology. Did your scientific research help shape your approach to your religious worldview as well?

This success at the Research and Development Forum for Science and Development was, of course, very important. The research has been underway for eight years now, from 1992 through 2000. It has been a very buoyant experience. And it seems likely to produce further such empirical results.

As my scientific understanding grew along with these successes, however, my questions became: What have I to do? What contribution can I make? Is it only in terms of research? If not, what else might it be? I went through a tremendous period of studying sacred scriptures for about four years. I focused on the Christian scriptures in more detail than any others. But I also explored other scriptures, including Hindu, Islamic, and Buddhist sacred texts. I also examined many of the philosophical texts of these traditions. The groundwork for this philosophical work was laid while I was in Cambridge. For two of my years there I undertook a course of study in the philosophy of science alongside my scientific training. So I had some inkling of where I should be looking. I also had the chance to do quite a bit of study in Europe.

The scientific successes you had led you to deeper questions of faith?

Undoubtedly, my science helped to guide my faith, just as my faith had motivated my science from a young age. But, to be honest, there was also a deep conflict that had bothered me for a very long time and that took years to resolve. It had to do with the demands of scientific methodology—what they include and what they exclude. Reading philosophy of science at Cambridge, I found some of the philosophical basis of science rather artificial. I had already started my research work well before I went to Cambridge, writing more than twenty papers before going there, so I had some practice to use as a touchstone. The scientific approach is both very powerful and very limiting: to look only at what can be quantified, what you can count or enumerate, and to consider only experimental evidence. You cannot use yourself as an experimental subject, because it would be difficult to differentiate yourself as an object. I found that very difficult.

Yet these tensions were powerful in your case—you didn't stop your research.

The scientific method is very powerful because other people can look at the same phenomenon and in principle replicate what you have seen or what you observed. Often they will even agree with you in the conclusions they draw—though their conclusions might differ depending on what was the precise hypothesis that both of you started off with. Still, I found the restrictions difficult because many of the ideas that I had, in my research, came through intuition and not directly from the facts I had gathered. In some cases I had gathered the facts and then looked at them to see how they fit with my hypothesis. But my greatest ideas seemed to have come from out of nowhere. In many cases the hypotheses were really very powerful, as subsequent experience showed.

The more I became conscious of this fact, the more I began to feel that the question of *intuitional revelation* should be integrated into a scientific approach. It is a factor that we scientists are forcefully trying to exclude; we tend to feel that the "lean" methodology of science has worked so well that we should not bring the more personal source of insight into it. Yet many times behind your results there is an intuition that is shared with others. Although we know this is true, the fear of, and restraint against, undue zeal continues to be a very important part of scientific work.

Yet, as you say, the greatest of ideas have come from this intuition. Is intuition something more than just the cognitive memory we now recognize in science?

That's the point. Intuition doesn't only live in your memory, not even transmemory, not even societal memory. Scientifically, one might look at it in this way: at the quantum level of reality the basis of material things reveals itself as nonmaterial—indeed, as transcendental. The process of observation by the experimenter creates reality. Thus elementary particles exist as waves when not observed and reveal themselves as particles when observed. Epistemic principles are to be regarded as rooted in nature, yet they do not appear in the visible order of nature. They are, instead, rooted in the transcendental dimension of nature.

Epistemology, cognition, intuition—all these are principles of the mind, just as the moral principles are principles of the mind. And the mind is a natural extension of what we might call the mindlike background of the universe. It is this mindlike background of the universe—by which I mean God—that gives science the authority to reveal knowledge. Otherwise, no facts, including the facts exposed by the so-called scientific method, would exist by themselves. Faith, it turns out, is essential to the process of deriving facts. Why? We have no experience of a cause-and-effect event; the principle of induction can hold only if the future resembles the past, and the permanence and identity of any object are only assumed.

Then what is intuitional revelation: Where does it come from, and how does it reveal itself to a scientist?

In the African religious traditions we say that intuitional revelation is a source of memory from the living dead. The living individuals maintain a constant, enduring relationship with the living dead. This belief explains our strong awareness of immortality and continuity and connectedness. It explains why we can never forget that life and death are not separate from science.

I see all of them together—the living and the dead, including all members of the extended family—connected and bound together into an intergenerational community. My life, in the African worldview, also includes the idea that we are connected to God. This belief gave me a totally new picture, a much wider picture than Christianity's view. There is a place where I am dead and I wait for Jesus to come back; I am resurrected, and then he can judge me. This connection between life and death, though, means that the distinction is not absolute; you are judged as you go along throughout life.

It is almost as though a thread of another form of existence runs through what we call life. Once again I hear the importance of interconnectedness in your view. Here also your views build on the unity of what it is to be human. How does that revelation—to use the term you used earlier—shape your scientific imagination?

I believe that both what it is to be human and creation itself are intimately linked with each other. African traditions teach the continuance of life. I attempt to think of life as continuous through all the various life forms. What we call life force itself can be regarded as the invisible, essential reality of everything that exists, living as well as inanimate—rather than making a distinction, for example, between plant life and animal life or between what scientists call single-cell beings and multicellular beings or social beings and nonsocial beings. Living beings possess more of the life force than the inanimate, and humans possess the most.

In the African view the living dead possess a great deal of the life force, but they do not have the physical bodies that are required to carry out actions as physical, material entities do. Thus one may think of the living human individual as existing in a physical, material universe in which he or she is linked intimately to *all* creation, both living and inanimate, through the life force.

This worldview seems to express the core of your belief about faith, science, and values. It emphasizes a respect for life and community and seeks an understanding of the unity of all of life with God through freedom, research, reflection, and action. I would like to end with this last question: What are your hopes for future research in science and faith—both for the scientific community and for the human community at large?

I would like to conclude with a thought, which is actually more of a feeling. I feel there is a force that runs through all these various forms of being: living beings, the living dead, and the inanimate. It seems to me that what we as scientists are seeking to find, and may eventually find, is what life means. In the end we will find, I think, a very elementary stage of life that runs through every life form and yet that it is not very specific for any particular being. If one were to start on a completely new form of research—and that's the direction I believe science is going—this is the question to which my research would be dedicated. I would ask: What is this force?

7. Faraneh Vargha-Khadem

Dr. Faraneh Vargha-Khadem was born in Tehran, Iran, and completed her graduate studies in 1979 at McGill University in Montreal and the University of Massachusetts–Amherst. After doing postdoctoral training at the Montreal Children's Hospital, she joined the Faculty of Neurology and Neurosurgery at McGill University, where she worked for two years before moving to London in 1983 to accept a faculty research position at the Institute of Child Health. There she created the first academic department of developmental cognitive neuroscience in the United Kingdom, along with its clinical counterpart, the Department of Neuropsychology at Great Ormond Street Hospital for Children.

Philip Clayton interviewed Faraneh Vargha-Khadem; Holly Vande Wall edited the interview for publication.

Vargha-Khadem's research and clinical work are directed toward understanding the cognitive and behavioral deficits of brain-injured children in terms of the underlying neuropathology, with the goal of developing new knowledge about the ontogeny of specific neural systems. With her colleagues she has made a series of landmark discoveries concerning the ontogenetic neural bases of episodic and semantic memory, speech and language, and differences in the functional organization of the developing brain as compared with that of the mature brain. In recognition of her contributions to the field, she was offered a personal chair in developmental cognitive neuroscience at University College London. In 1998 she was elected a Fellow of the Royal Society of Arts and in 2000 a Fellow of the Academy of Medical Sciences. Vargha-Khadem is a member of the Baha'i Faith, a world religion dedicated to establishing peace and unity among humankind.

Faraneh Vargha-Khadem speaks with a cultured British accent, with just the slightest touch of Farsi discernible in the background. She blends the sharp, analytic mind of a researcher in the neurosciences with a deep compassion for her patients; one imagines her not only running a research team but also sitting at the bedside of seriously sick children. One has the sense of conversing with a cosmopolitan scholar, a person whose interests and allegiance are not tied to a single land or people. Discussions move seamlessly from neurological details to the history of religions, as if no effort were required to hold together such diverse realms of human experience in a single, harmonious whole.

Can you describe the events that led you, as a neuroscientist, to specialize in memory and in amnesia in particular? Was that an interest from the beginning of your graduate work, or did it evolve in the course of other research?

My interests in amnesia are actually the result of a personal evolution in interest. In the earlier phases of my research I was predominantly involved in studying speech and language functions in children with focal brain injuries. At first the work on memory was secondary to the work on speech and language. But when we discovered a group of children with the symptoms of developmental amnesia, my interest in this topic took over.

Can you describe what happened?

About seven or eight years ago my colleague Mortimer Mishkin and I started seeing a patient who had been referred to us because his problems could not be readily explained by the other doctors who had seen him. In our investigations of him we found out that as a result of a pineal gland tumor, he had developed three very rare conditions during childhood. One of his problems was that he was completely unable to read subsequent to the discovery of the tumor, although he had developed reading and writing skills to a very high level prior to the diagnosis of his tumor. He had essentially become alexic. Simultaneously, he had also developed visual problems and was unable to recognize objects that he encountered in his environment. Finally, he had developed amnesia. So he had three syndromes in one. We started investigating this young man and found out that his amnesia was of a unique nature; no one had ever documented this particular form of amnesia. He could recollect all the events and episodes that had happened to him prior to the onset of his tumor, but subsequent memories were not accessible to him through *oral* recall, only through writing. He could actually write his memories, but he didn't know what he had written—or, more accurately, he had no conscious awareness of what his hand had written—we had to read back to him what he had recollected; he could not have read it for himself anyway, even if he had had the awareness.

This case essentially became the beginning of my work on amnesia. The paper that we published on him came out in 1994. Since then we have pursued this area of research, investigating the very selective nature of memory impairment that can be found in children, that is, cases where the pathology is sustained very early in life, before any memory development has taken place.

Can you summarize the findings of the past several years and some of the insights you've gained from your work?

Well, essentially we have discovered a new syndrome, which we have called "developmental amnesia." This syndrome can result from selective brain damage occurring very early in life. In some ways developmental amnesia is similar to adult-onset amnesia. For example, as in adults there is a profound impairment in episodic memory, that is, in the ability to recollect events and episodes of one's life. This type of ability is heavily impaired in the group of young patients that we have been studying. Contrary to adult amnesics, though, who have an impairment in semantic memory as well (that is, in the memory of facts and general knowledge), these children have developed their semantic memory very well. They have learned to speak, to read, and to write, and they can attend mainstream schools and acquire general knowledge. More important, they have developed fairly normal levels of intellectual ability. So in many ways they seem to be normal, and yet they suffer from crucial abnormalities. They lack one of the essential hallmarks of independent living: the ability to remember events in one's own life.

What is responsible for the deficit?

The pathology that appears to be responsible for this syndrome is atrophy of the hippocampus on both sides. It seems to result from hypoxic ischemic episodes, or events that deprive the brain of oxygen. All the children with developmental amnesia that we have studied so far have suffered from such episodes, sometimes as early as the first few hours of life. The hypoxic ischemic events may be associated with difficult births or extreme prematurity or with other medical conditions such as prolonged seizures. But despite the fact that these children were young at the time of their injury and had all the plasticity of an immature brain available to them, they were nevertheless unable to overcome the consequences of the brain damage, at least in terms of rescuing the capacity for episodic memory function.

Two aspects of these results would be fascinating to pursue. The first is the ability of the brain to compensate to such a remarkable degree—something these children show but that one doesn't find in cases of adult-onset amnesia. The second concerns the limits on compensation due to the pathology. Were the results surprising—the degree to which they were able to make these compensations and the limits on what their brains could do to compensate?

Well, yes, they were surprising. We are used to seeing dissociations in adults, with [for example] preserved intelligence and yet impaired memory and learning. In these children, though, there were not only those sorts of dissociations *between* cognitive domains but also *within* the domain of memory specifically.

Are these new findings?

The discovery of such dissociations is most certainly new. The other interesting aspect of the problem is that the deficient parts of memory are not obvious at first sight and are veiled by those aspects of memory that are normal and well preserved. The first three cases that we saw were referred through the medicolegal system with the query of whether there really was a genuine memory problem that could explain the difficulties of the children. Although the parents and teachers of these children all complained that there seemed to be a problem with memory, they could not be certain that the problem was genuine. They wondered whether the children were using the excuse of forgetfulness to hide their laziness or lack of motivation—maybe that was why they were behaving in unorganized and irresponsible ways. Because of this we had to first determine whether there was a personality disorder or a true learning disability.

If even the parents and teachers were unsure, the children must really have been able to compensate remarkably for their memory deficits.

That's correct, and of course that made our job much more difficult. Our first task, then, was to actually determine whether we were dealing with a genuine neurological problem or not. Then it gets more complex: once we realized that the memory dissociative disorders were indeed genuine, then we had to look for causes and determine their origin. After the children had overcome the stormy events surrounding their births, they usually did not have any further serious medical conditions that could have produced such a selective type of brain damage. So it was difficult to discern how the brain damage could have occurred in the first place, since the children did not show any other impairments except the one in memory. More important, they did not have any motor impairment.

Was the actual brain damage difficult to locate?

It was initially, because although routine magnetic resonance imaging can show the presence of abnormalities, it does not allow any quantitative measures of the extent of damage, in this case, hippocampal damage. Thanks to the development of quantified magnetic resonance techniques by my colleague David Gadian and his team, we were able to measure the extent of hippocampal damage that was needed to produce the amnesic syndrome. We could then draw links between the extent of hippocampal damage and memory impairment.

What theoretical framework do you use to connect the physiological damage with the actual cognitive performance of these patients?

The framework we use grew out of research with monkeys carried out by Mortimer Mishkin and his team in the Laboratory of Neuropsychology at NIMH [National Institute of Mental Health]. Mishkin has developed a theoretical model of human and nonhuman memory development based on the physiology of the hippocampal system, his data from his work on monkeys, and the data that we have gleaned from our own memory research. The model is different from the ones that have been around for some time insofar as it proposes that the memory system is hierarchically organized, with the hippocampus sitting at the apex. When streams of stimulus traces from the sensory modalities are fed into the hippocampus via the subhippocampal areas, they become more and more enriched with spatial and temporal features as they travel up the hierarchy of the hippocampal complex, making each event and episode of our lives unique. So if there is damage in the hippocampus, which sits at the top of the hierarchy, then necessarily the information that gets fed into the system is going to be degraded and devoid of the spatial and temporal features that define each episode and make it unique in our memory.

In this view the hippocampus serves as the center that structures and enlivens our memories as they are relayed through the senses. Normally, the hierarchy is set up sometime during early childhood and becomes more fully functional with increasing age and experience. This would explain why very young children cannot often fully recall events or episodes in their lives. But if the hippocampus is damaged, recollective abilities may not develop with increasing age and experience and may remain very immature and primitive. Damage of the kind that we have documented directly affects the ability of the hippocampus to give spatial and temporal texture to life events.

How is it, then, that children with developmental amnesia are able to compensate? How is it possible for them to show such cognitive ability when there is significant damage to the hippocampus?

Hippocampal functioning emerges very early in life, but it may not be fully functional from the start because different regions of the brain may not come "online" uniformly during the developmental process. The structures are in place from the moment that the baby is born, and obviously even during the gestational period these structures are developing, but it is the experience that the child has in interaction with its environment that's going to actually bring out the functional specialization of these systems. Complex interaction with a child's environment yields the functional specialization of the brain, especially in the instance of structures of memory formation. So in the absence of a normal hippocampus—even when it's damaged early in life, when there is the greatest extent of plasticity and reorganizational capacity available—it is still difficult for the young brain to compensate for the damage to this structure and to develop normal episodic memory. However, provided the damage is restricted to the hippocampus and does not affect the cortex underlying it, then other aspects of memory such as semantic memory, which is not heavily dependent on hippocampal processing, can be preserved and can develop normally. If the damage is compartmentalized, so to speak, as in the case of our studies, then other forms of memory can develop normally. So according to the model we employed, the near-normal semantic memory seen in children with developmental amnesia is possible because the cortex lying subjacent to the hippocampus is intact. This cortex is capable of processing semantic information—which is factual and is not set within temporal or spatial contexts—but it cannot accommodate the storage or the processing of context-rich episodic information, such as spatialized and temporalized memories. These fall within the domain of the hippocampus. Since in these children both hippocampi are damaged as a result of anoxia, there is no other structure that can take the place and do the job of the hippocampus.

Interesting. So if I understand correctly, much of the theory that you have just described is new? That is, these are new areas of hypothesis and research that weren't in the literature prior to your work?

In fact, there were no cases of amnesia reported in children below the age of eight. And it was thought that there could be two possible explanations for the rarity of this condition. Some believed that if this form of selective hippocampal damage occurred early in life, the child would just become mentally retarded rather than amnesic, because memory is so vital to early childhood development. Others believed that, because such trauma occurred when the brain was most plastic in its functionalization, then hippocampal pathology would not really show an effect because the consequences of the damage would be overwritten by compensation and rescuing operations elsewhere in the brain. But the cases that we have reported argue against both explanations, as we've been able to document instances of children with fairly normal development who nonetheless lack the ability to retain memories.

What implications might this have for a broader theory of memory functioning?

It's fair to say that certain theories that attempt to explain how memory function takes place and how the hippocampal system works have been questioned now as a result of the dissociations we have found in children with developmental amnesia. There are several theories on the role of the hippocampus in memory, such as the influential theory proposed by Larry Squire, who does not view the hippocampal system as a hierarchy. He sees semantic memory and episodic memory as different cognitive processes that are subserved by the same neural system in the medial temporal lobe. Episodic and semantic memory are essentially unified in the medial temporal lobe for Squire. According to Squire's view of a unified medial temporal lobe system, a dissociation between semantic and episodic memory cannot result from selective hippocampal pathology, since it would impair both episodic and semantic memory equally. In contrast, Endel Tulving, another influential figure in the field who has written extensively about the cognitive processes, distinguishing episodic from semantic memory, agrees with the hierarchical model that we have proposed because it provides a neural basis for the well-documented dissociations in cognitive memory. So our model proposing the hierarchical organization of memory has drawn many neuroscientists into a debate. A recent issue of the journal *Hippocampus* was dedicated to these different points of view, as represented by Larry Squire, Stuart Zola, and colleagues;

Endel Tulving and his colleagues; and Mortimer Mishkin, David Gadian, and myself.

So in this discussion you've begun to work out some of the cognitive and theoretical implications of your anatomical discoveries. Have you also begun to make connections with other disorders employing your model?

In order to make further connections with other potential cognitive disorders, we've been working with Alan Baddeley, who is well known for his work on short-term memory. At first Alan was skeptical, like Larry Squire, of the hierarchical model of hippocampal functioning. But now, having tested some of our patients with developmental amnesia, he accepts that these aspects of memory can dissociate following selective hippocampal lesions. With his help we have now found a second dissociation, this time between recognition and recall, in the same group of children with developmental amnesia. Alan has always been interested in finding out whether recognition and recall could dissociate as a result of different types of brain damage. After years of looking for this dissociation in different populations of patients, from Down syndrome to those with temporal lobectomy, et cetera, he had finally given up, thinking that such a dissociation does not occur. His investigations led him to conclude that if there was an impairment in recall, there would be an impairment in recognition as well. That's when we asked him to see one of our patients with developmental amnesia, who showed a very severe impairment in recall, along with very well-preserved recognition ability. These behaviors suggest that this patient's hippocampal lesion has blocked recall but that his recognition ability may be located elsewhere, possibly in the domain of semantic memory processing. So our work with Alan has allowed us to make further connections using our previous work on hippocampal legions and the model we employed in that research.

Can we make the transition to religious questions? Before I ask what kind of connections you draw between your neuroscientific work and your religious interests, though, could you briefly explain your religious background?

I've had a very mixed background religiously, as I grew up in Iran—a Muslim country—was sent to a Catholic school as a little girl, and was raised in a Baha'i family. My strongest affinities, though, are with the

Baha'i Faith. My great-grandfather became a Baha'i 157 years ago when the Baha'i Faith was born, so it's been a very strong influence that has shaped not only my own development but certainly that of my family as well. It's within the Baha'i tradition and the framework that it provides—even though it's a very young tradition—that I have found unity between my scientific work and my spiritual inclinations.

Many scientists speak of that unity as difficult to achieve. Did you experience tension between the approaches and the mind-set required of you as a post-graduate student and your religious tradition, or was there a sense of natural integration for you throughout the course of your studies?

For me personally there wasn't so much of a tension, though for many people with whom I came into contact there was more of a tension, as many of them were agnostic or atheists. And many of them had a great influence on me. It was often difficult to reconcile my acceptance of their views on neuroscience and yet not to accept their agnosticism. I never felt such a tension, however, probably as a result of my Baha'i faith. I realize that science is limited, inherently limited, because it emphasizes quantification and definition. Science can only deliberate over what is quantifiable or measurable, and I think this makes up only a small part of our sensory or intuitive experience—especially when it comes to encounters with other people. In this spirit I think that we have to see science as a tool and a means, one that can take us only so far. We should use it insofar as it allows us to understand tangible and material truths. But this doesn't mean that science exhausts all of experience. Hardly. If you see science as a tool, then the conflict between science and religion doesn't necessarily go away, but it's certainly much more flexible.

Science involves a particular mode of knowing, and you've described some of its attributes: quantification, the empirical method, and so forth. But there are other modes of knowing, which you spoke of a moment ago as intuitive, that are more naturally characteristic of religious or spiritual traditions. How would you describe them? If you had an agnostic friend who asked, "Tell me about these other modes of knowing. I'm interested but skeptical about them," how would you respond?

It's difficult to convey what intuitive knowing is or even what it feels like. I think that there is a common understanding of intuition that can be discussed, especially among more sensitive people. You can give examples of instances of intuition, and I guess you can talk about them. Although you can never prove that a specific type of awareness is present, you can have some understanding of it and can speak about common experiences. The older I've become, the more my friends and colleagues are willing to discuss such phenomena, inexplicable as they are. Maybe it's an acceptance and a willingness that comes with age. When we are in our twenties or thirties, perhaps we want everything in very concrete forms, and we are uncomfortable when we cannot get complete explanations—young people leave very little room for uncertainty. But as we get older, I think we recognize that there are many layers of complexity, and perhaps there are many things in life that we will never be able to explain fully. But that doesn't mean that they don't exist or that we will never be able to comprehend them. Rather, it means that we have to be willing to accept the uncertainty that comes with talking about such things. Perhaps this acceptance of uncertainty comes with age, with reconciling ourselves to our mortality. Recognizing it allows us to accept the uncertainty of, say, talking about intuition and so forth.

Yes, increasing age and facing one's own death raise questions that seem impossible to answer within scientific inquiry. They are of a broader nature.

Yes, and these questions are not a matter of proof at all. It's a matter of belief, a matter of feeling, a matter of knowing inside yourself. But that doesn't mean that one cannot have these discussions, and it doesn't mean that one can't try and put oneself into the position of the other, just to see what it's like to look at the world from that perspective. Just because these questions are beyond the view of science doesn't make them indescribable or uncommon. They address what is actually *most* common among us—this thirst for belief, for answers to the unquantifiable questions. Somehow, the majority of the people of the world seem to have a need within them to aspire to and to understand things that are beyond themselves. This appears to be such a primitive and instinctive need that almost everyone manifests it in one form or another, from the most primitive cultures to the most sophisticated ones. From this perspective it's actually the desire for quantification and tangible knowing that is unique and different.

And do you see this common desire for, and experience of, the unknown as what unites us as well as dividing us?

I know that a lot of people think that the wars and the difficulties we have in the world are all due to religious conflict. But I don't see it that way. I think the conflict between adherents of different religions is man-made. Religious philosophies aren't exclusive of each other or in direct conflict with one another. In essence, all religious philosophies are the same. All are attempting to bring a prescription for harmonious living to the people of the world. If this message becomes perverted in practice or within institutions, then that's the fault of flawed human beings, not of the doctrines themselves. But the fundamental attempt of all religions is to address these flaws and to fashion out of them a type of harmonious living.

I hear reflected in your answer two kinds of harmony. Baha'i clearly asserts harmony between science and religion, though you describe it as a complementarity, not a unity. At the same time you also describe a harmony between the world's religions, a unified message at the most fundamental level. At this deepest level do you believe that there is a single religious or spiritual message that all or most of the world's religious traditions are expressing?

As a Baha'i, I feel that I'm as much of a Muslim or a Christian or as much of a Jewish person as any other, since fundamentally all these religions are teaching the same truth. It's really in virtue of the time in history that these religions originated, and the revision of the social laws that they have brought with them, that they are different. The revelation for each emerges out of a specific social and historical context. But the essential teaching—which is to know God and to reflect the qualities of God, which are all good—is the same, and it unites each religious philosophy. The fundamental teaching is always one of harmony, of relating one's historical situation to the reality of an all-good god. None of the religions prescribes anything different. As Baha'is we believe that there is progressive revelation, and that each religion, by virtue of the fact that it comes later and later in history, is more advanced and more modern so as to suit the needs of mankind. There is a constant progression of revelation; it just so happens that the Baha'i Faith is the latest chapter in this ever-progressive process. Every religion has a cycle of approximately one thousand years during which its goals are achieved. And once that cycle is completed, there is need for the renewal of the civilization that it inspired. This renewal doesn't mean that the essential spiritual principles of the previous religion are going to be discarded or outdated. It simply implies that the teachings of any religious philosophy and its adherents need to be adapted to a new historical or social situation, one that the old teachings have ultimately helped to create.

So in your view many of the differences that outsiders see among religions really reflect cultural and historical differences, not fundamental differences?

That's right. For example, when Christ came, his message was one of unity for small groups of people who perhaps existed in villages or lived in small tribes and clans. And when Muhammad came, it was from him that the idea of a nation, the Arab nation, was born. So there emerged a national identity associated with the movement of Islam. And with the Baha'i Faith the fundamental goal is one of global unity. There is a continual expansion of those to whom religion applies. It's an awareness of people as world citizens. Along with this idea have come basic principles on how the world could be united, so that it would indeed be as one entity without artificial divisions or unnatural borders separating people on the basis of national or racial identity. This doesn't mean that Baha'is don't recognize the distinct linguistic and cultural traditions among us all or that these should not exist. All these variations should be safeguarded, of course, because they are reflections of the collective experiences of groups of people. But, fundamentally, all religions are equal, made by the same Creator, and capable of generating harmonious living among all human beings. The differences between human groups are not fundamental but always point to the common experience of attempting to create community—a community that lives harmoniously with itself and others.

Could we talk a little bit more about some of the connections or complementarity between science and religion? Based on your knowledge of the neuro- and cognitive sciences, are there any indications that point to this spiritual side of humanity and hence might yield a greater understanding of who we are?

This is where I believe that science is too restrictive, too narrow in its focus. Nothing tantamount to a spiritual dimension is allowed to emerge from scientific inquiry. Just because the spiritual realm cannot be defined according to scientific principles, science seems to have decided that it doesn't exist, so this realm has been excluded. The criteria of strict definition and quantification have forced science to rule out what most of us find to be quite common. I don't think the reverse holds true, though. I don't think that religious philosophy denies science. Certainly, in the Baha'i Faith this is not the case. Most religions allow for something like a scientific dimension to express itself in humanity.

Can you think of ways in which religion allows for the expression of the scientific spirit or shows how something like science could come to be?

The Baha'i writings describe the many ways in which we human beings share our primary senses with the animal kingdom. In this respect we are no different from animals, and I think some of our social behavior reveals that we can often be animalistic. However, we differ in one significant respect: human beings have a soul and animals do not. As a result of our soul we have a number of senses and potentialities that animals do not. These are described in the Baha'i writings as the inner powers that are energized by the soul. These powers express themselves through the sensory modalities. For example, just as energy can be seen as light reflected through a lamp, so the soul can channel its inner powers through one or all of its sensory modalities. There are five inner powers: memory, comprehension, thought, imagination, and the common faculty. If we translate these Baha'i concepts of inner powers into everyday language, it appears to me that they are precisely the domains that cognitive neuroscientists are now investigating. Neuroscientists are studying perception, language, memory, thought, and consciousness, and they are of course looking at the brain to see how these processes are represented.

And as a religious scientist you see yourself as engaged in the inquiry into these various faculties of the soul?

As a neuroscientist I am deeply interested in examining the same processes and modalities described in Baha'i writings. I would also like to learn about how memories are formed and recollected, how perceptions are created, et cetera. By analyzing brain function, or by analyzing the memory system or the language system or the neural substrates that facilitate them, we can better understand how these processes take place in the brain. Yet I'm convinced that we will never understand the *source* of these processes. Going back to the example of the lamp emitting light energy, I would be deluding myself if I thought that by studying the lamp I had understood the source of the light energy. And so by studying the brain I can hope to understand how memories are formed or how language is produced or how perceptions are formed through the activities of networks of cell assemblies. But I will never be able to determine the origin of those cell assemblies or networks, much less why they exist. At least not within the realm of science. Science doesn't allow me to answer such questions or really even to ask them. Each memory or spoken word or percept or thought is peculiar to the individual and thus is an expression of that individual's soul or human spirit. And it is unique, just like my power of perception is probably different from yours. And it's unique to me, because it is my soul that allows me to perceive the things I see in my environment. My investigations as a scientist merely touch the surface of what is actually occurring in the human being, given the inexplicable power of the soul.

So there is a fundamental difference between every two human beings, because each has a unique soul? Does that mean that we could never have an overarching theory of, say, memory, one that could be 100 percent adequate in explaining your particular memories or my process of memory?

All of us have unique senses that are driven by our human spirit. Maybe some of us have stronger powers than others in certain domains, just like our outer attributes vary from one person to another, such as intelligence, beauty, communication ability, et cetera. For this reason I don't think an absolute explanation of memory or even perception can ever be formulated. We can surmise general explanations that obtain in most cases, but we can never fully or reductively explain an individual's memory or perception.

How do you respond to the attitude that one often encounters among neuroscientists, which is that before too long we will overcome these limitations? They argue that, as our imaging techniques and our theories get better and better, the adequacy of our understanding of memory and perception and so forth will continue to increase, until some day an empirical science—some combination of cognitive neuroscience and genetics and so forth—will provide a full and complete understanding of human cognitive functioning.

I honestly don't think that a fully reductive explanation of mental processes is possible. We may be able to gain a greater knowledge of the brain's physiology and chemistry, but, because of our ensouled natures, I don't think we'll ever fully grasp the "light within the vessel" that makes the mind work. We can identify structures and relationships but not causality. Science is inadequate for fully comprehending these matters.

That means that there is a dimension of what it means to be human that forever recedes from the scientific quest?

Unless we discover a science that transcends empirical evidence, there will always be a dimension of humanity that is unavailable to scientific explanation. This is why it is enlightening to read some of the conclusions that the great thinkers of our time have come to—those who really have left the door open on the question of the existence of God, the existence of a world beyond. They've left the door open because they have recognized that the human mind can go only so far. They've realized that human understanding always comes to an impasse. The only thing keeping us from a deeper understanding, of course, is our finite minds. But I don't foresee our finitude changing any time soon—certainly not in my lifetime. Maybe in the next cycle of civilization …

Are there sciences or groups of sciences that could come closer to an understanding of the spiritual nature of reality? For instance, is physics less capable than evolutionary biology, which at least understands the notion of progressive change, or is biology less far along than cognitive psychology, which can enunciate principles of purpose, will, ethics, value, and so forth?

I personally think that it is very much a function of the individual. Interestingly, as a student I often encountered evolutionary biologists, whom one would think would be quite open to the questions of belief and spiritual phenomena, who were either indifferent or hostile to such questions. And, ironically, I've found physicists and computer scientists often quite receptive to that line of inquiry. So, as I said, it appears to be more individual than anything, and, interestingly, that kind of inquisitiveness might be independent of whatever form of scientific training someone has.

Do you believe that these differences are also a function of what you referred to earlier as our individual spirit?

Just as memory or perception is individual and cultivated, I believe that one's spiritual understanding is individual and can vary significantly. Through either natural ability or learning, one can attain higher levels of spiritual understanding. Interestingly, it is this sense of spiritual learning or cultivation that is lacking in childhood education. We nurture children's physical and mental development, but we neglect their spiritual education. If the child's upbringing has involved spiritual nourishment and nurturing, then this will be maintained not only throughout development but also through adulthood, like all other attributes and knowledge that have been acquired. But if the child hasn't been taught to cultivate this interest, or even to look for spiritual phenomena, then that aspect of its being may not ever be awakened. But of course there is always a substrate, there is always a predisposition toward spirituality in humans. It just depends on what one does with this predisposition.

I have been asking scientists about what it is like for them as religious individuals to be involved in their profession. Are there any ways that you find that your religious belief and practice actually assist you in doing your research, working with subjects, and formulating theories? Or are the two sides more separate than that?

Well, they are a little separate, I'm afraid. It is only when I go beyond science and ponder and reflect deeply on the implications of some of our research findings that I find unity between my religious convictions and scientific work. For example, another aspect of our research work is on speech and language, investigating the genetic basis of speech. Some of the geneticist colleagues we are working with are now very close to identifying the gene that appears to control the ability to speak. It is absolutely remarkable to think that the brain of human beings might be put together such that speech is produced as a result of the regulatory function of one particular gene that controls the ability to produce the sounds making up intelligible speech. And though this gene is found in almost all humans, it is still completely unique in its expression, as we discussed earlier. It is incorporated into the human genome as a functionally specialized gene, and yet it allows for incredible linguistic diversity. It's at such points, when I am fascinated by the complexity and diversity of concepts like these, that a realm beyond scientific inquiry becomes quite evident. For at these points I realize that I can seek only to understand how such a system works, not the *why* of it's being there ...

Wait, can I just make sure that I understood the last point? Are you saying that such a remarkable result, such as the discovery of the gene responsible for speech, gives you a sense of a broader purpose, a structuring of the universe for which God is responsible, that we can only begin to grasp in our research?

Such a reaction is only natural, and I think it complies with the reaction of many scientists when they witness something of both inordinate complexity and purpose. But I have to be cautious about employing such explanations in my work, because I work with children, very severely brain-damaged children, and if I am to help them with their disabilities, then I can only remain within the realm of present-day science and hope to make a difference one increment at a time. For example, we have recently been working with a group of children who have sustained frontal lobe injuries. When they are developing through childhood, we don't see the damaging consequences of this type of brain pathology on social behavior. But as these children mature and become teenagers, they gradually become sociopathic, a threat to themselves and a menace to society. So now the whole question of abnormal ethical and moral behavior has become linked to brain pathology, suggesting that there is even a biologically programmed sequence and neural substrate for appropriate social and moral behavior. In this instance I have to confine my reasoning to scientific explanations while often setting aside or making secondary some of my religious convictions.

Can you explain how religious people might be tempted to respond differently to such cases? We're talking about people who can't be integrated into society, who are somehow dangerous to society. Would a religious person be inclined to see them differently?

I do see them differently, because they are patients and as such need help. I think this is a conviction informed by my belief in the value of the human soul. And, second, I don't think that their sociopathy is a reflection of any spiritual shortcoming but rather of a mental or biological shortcoming. Perhaps if these patients were diagnosed early in life and identified as being at risk for later sociopathy, they could undergo preventative treatment so that their social behaviors would become appropriate through early education. Children typically learn moral behaviors through osmosis, through observation, but if there is damage to the regions of the brain where social behavior is regulated, then there is no appropriate neural substrate for this behavior to emerge normally. In such cases what is likely needed is systematic education in social behavior. And while these are fairly normal scientific decisions, my empathy for these patients is formed through my religious belief.

A broader question about understanding the human person has, in the past, been approached in very distinct ways by scientists and by religious people. Religious people would draw on their particular tradition and its scriptures; scientists, of course, on the empirical method. Are we at a point in human history where those two aspects might now work together better to help humanity understand itself—to understand what it is to be a human person in a broader, fuller way?

I think such a cooperation has to occur, especially as we all become more and more integrated globally. It is no longer possible to see solutions to economic and political problems purely through the lens of science or the material realm. We have to constantly look at the global picture, to seek solutions that are informed not only by science but by religious and cultural values as well. And I think we are already seeing such a transformation. As we have become increasingly interrelated, we are seeing more and more opinions being expressed where there is a unity, or at least the hope for such a unity, between science and a modern religious philosophy. For example, not too long ago there was an editorial in the *International Herald Tribune* about how science, when guided by ethics, could lift up the poor. I think we will begin to see more of these types of pronouncements, and in fact I believe that they're necessary if we are to find solutions to the critical problems that face us.

You are a specialist in neuroscience and cognitive psychology, and a believer in a religious tradition that has a unitary vision of the world's spiritual traditions. From that dual background and dual set of resources, can you briefly describe how you see the human person and our place in the universe?

I think that each person in his or her own way helps advance civilization. Even though we may seem very small in the great overall picture, we serve an important role, each in our own way. And we do this through our work, which should be put to the service of humanity. I think our purpose in life is to raise a new generation with the awareness that their efforts and contributions should not be only for themselves and their families but also for the betterment of humanity at large. I try to raise my own children to become world-minded, rather than inward looking, so that whatever they undertake in life will benefit many people. It doesn't matter what they do, as long as what they do is actually something that will help others. Perhaps if this were the goal for the majority, then the world would be a better place anyway, and it would leave long-lasting results.

So you see your role as not only for the present but also projecting into the future?

One of the things that we as human beings have is that we can look to the future, we can see beyond ourselves, and we can see how our actions ripple through the course of time. It's an incredibly important part of a human being's life—maybe the most important part—to be able to leave something for the future. They say that it's for your children, but in fact if you really think about it, it's really for future generations. The ability to conceive that what one does now will in one hundred years benefit others is something that is unique to human beings, and I think that it is spiritual. It reveals how interrelated we are through time, how my actions that advance civilization now may also be operative in the future. It reveals how much gravity is attached to our actions. Knowing this does not satisfy you immediately—it is a motivating force, though, one that keeps us striving to leave our mark or to leave children behind who will carry our philosophy forward. We realize, sometimes only implicitly, that what we do now inevitably touches on the future.

That's a beautiful answer to the question. Does your belief in God make you more optimistic about this future?

I think the blueprint for progression is there, in all of us. Even if we were to have terrible calamities happening to the world during our lifetime, I think that out of it will emerge something good. While pain and suffering are unfortunate and horrible, I'm sure a greater civilization will be born.

8. Pauline Rudd

Pauline Mary Rudd is a noted biochemist whose work has made important contributions to the understanding of the immune system and to the development of therapeutic drugs for viral and autoimmune diseases. She is also a thoughtful Christian in the Anglican tradition—a tradition she embodies in a deep contemplative and experiential spirituality significantly influenced by Kierkegaard, Tillich, Hesse, and Eastern thought.

After earning her baccalaureate degree in chemistry at Westfield College, University of London, in 1964, Rudd served for five years as founding scientist and senior chemist at Wessex Biochemicals/Sigma in London, where she was responsible for research, development, and production of sugars and sugar phosphates. While

---

Philip Clayton interviewed Pauline Rudd; Jim Schaal edited the interview for publication.

raising four children with her husband, an Anglican parish priest, she continued to work part time in research, teaching, and consulting. In 1981 she returned to work full time and in 1982 completed her doctorate in chemistry at the Open University, Milton Keynes, where she studied the structure and function of glycoforms. She joined the Glycobiology Institute of the Department of Biochemistry at the University of Oxford in 1983, under the direction of Raymond Dwek. Rudd is now University Reader in Glycobiology at Oxford and glycoimmunology group leader at the Glycobiology Institute, where she heads a large research team focusing on glycosylation in the immune system. Her research—much of it carried out through international collaborations—sheds light on the critical role of sugar complexes in protein molecules that serve as antibodies. The results find wide application to diseases such as AIDS, heart disease, hepatitis, and arthritis.

Biochemistry is only one facet of Rudd's life. A dedicated mother, Rudd remains close to her adult children and sees parenting as a lifetime calling. A long-time associate of the Anglican community of St. Mary the Virgin in Wantage, Oxfordshire, Rudd also holds a strong sense of religious vocation as a layperson. She sees her three callings—to science, faith, and motherhood—as inextricably intertwined. This strong sense of integration, as well as her contemplative perspective, carries over into her extensive speaking and writing on science and religion.

In person Rudd is exceptionally warm and open to conversation. Speaking in a soft voice inflected with an Oxford accent, she moves effortlessly between personal anecdotes and literary allusions. At times voluble, she seems equally comfortable with reflective pauses before responding. Throughout the conversation one senses a spirit of intuitive exploration, gentle companionship, and a quiet appreciation for mystery—all central features of her approach to the divine.

"Perhaps some answers lie in the idea of partnership or dance, with a role for humans to work co-creatively with God and nature. Partnership, like dance, is not just one sided," she notes. "It involves each partner's putting the other first, respecting the other, relinquishing total control and extending to the other the trust and freedom to make informed choices. We cannot control God, and if we wish to be responsible individuals, then we cannot wish that God should control us, either. As we mature spiritually, we recognize that we are involved in a gracious, living, evolving partnership in which both we and God, or at least our perception of God, might be changed in the encounter."

Was a religious tradition an important part of your own spiritual formation?

I grew up in the Anglican Church, the Church of England, and it certainly was formative. It was the culture that gave me the language to express what I experienced. When I was a very small child, I began to learn and use religious symbolism and concepts to express my experiences. It's always been an integral part of me, and as I've grown older I've developed the use of the language. I haven't had any formal theological training, but I just learned within the context of a congregation—reading, discussing, writing, and listening to people—personally rather than formally. I have also been very fortunate to interact with some exceptional spiritual directors.

As you began studying at university and doing more advanced work in biology, did you experience conflicts with your religious beliefs?

I've never experienced conflicts. I never really accepted anything unless it correlated with my own understanding, so I didn't feel obliged to hold to any dogma that didn't seem relevant or fit with my present experience. As I grew, I simply evolved my religious practices to cope with my growing needs. Take the Apostles' Creed, for example. I remember as a teenager having real trouble with the idea that Jesus descended into hell. The definition of hell was "where God was not." I reasoned that when Jesus descended there, God was there—so the statement was irrational. I simply didn't say that bit in the [Apostles'] Creed for a while.

Later I understood that religion evolves and that many of the beliefs we have in Christianity are very ancient. The Christian idea of sacramental bread and wine goes back to the harvest gods of Neolithic man. To me that's what makes it special—it's as if religion has evolved with human thinking. This is part of the inspiration of the Holy Spirit: the ability of each generation and each person to reinterpret anew and to test what they believe.

What's really important is to hold to your own integrity. If a commonly held belief allows you to say, "This helps me to understand what's going on in my life," then that's fine. But if it doesn't, the worst thing you can do is pretend that it does.

As you describe it, your attitude toward your religious belief is somewhat similar to the attitude of a scientist, is it not—a scientist who sifts the data, accepts what she finds good reason to accept, and exercises flexibility about traditional theories where they no longer fit neatly?

I think God is very big. No one person, no one tradition, has it all together. What I'm interested in is the truth, the real answer. In science I don't have a preconceived idea of what I want an experiment to show. Instead, I want to look at the data and then ask, "What is it telling me?" I suppose it's exactly the same in my religious life.

If I don't understand or can't come to terms with something—either scientifically or by religious experience—I simply leave it as unresolved till I have more life experience or till it somehow becomes clear. Often that does happen. There are some ideas in Christianity that I still don't understand or that don't seem important to me. But as you go through life, any religion is like a vast treasure house. You're like a child on a beach picking up a pebble; at some point later in your life an idea may suddenly come into focus. For example, take the suffering of Jesus on the cross. I found this very hard to understand until I actually came to a moment when I suffered personally. Then it helped to know that God was suffering alongside me.

Some things in religion don't seem particularly rational or relevant, until I come to the moment when I need to understand. Then I say things like, "Now I understand what Jesus (or this psalmist or that prophet) meant, because now I know it for myself." The insight expressed in a religious text or artwork comes into focus for me when I identify with the experience of the author or the artist. Then the insight comes alive for me. I think, "Isn't it wonderful? I'm not alone, there's somebody else who experienced what is happening to me."

Is this where the analogy to the scientist's practice breaks down? The religious insight comes alive in an individual, existential moment. Is it really the same for scientific insight? Or is there a distinction here between subjective and objective insight?

To some extent, you do science day in and day out, and you need to progress one way or another. But I do think the things that really make me excited in science also spring from experience at a particular time.

I work with glycoproteins, which are very beautiful molecules. When I'm really familiar with a molecule, I've always felt that I'm actually inside it, that I can walk around it. It's huge and noisy and multicolored. In a sense, I understand how it feels, and I know what it'll do if I put it into acid. Wandering about molecules gives rise for me to a sort of inspired wondering: there's always something about them that's still a mystery, that I won't understand until it's revealed by more data. That sense reflects a close similarity to the way I feel about spiritual things.

Scientists sometimes talk about "inhabiting the world of their theories," having an intuition for that world. Is that what you're describing?

Yes, like Kekule, who dreamed the structure of the benzene ring. But the intuition has to be rational as well. In science the intuitive may give you a big leap, but then you have to back it up with experiments that prove your intuitive leap had some justification. I think that is true of religion too. In religion your intuition may leap ahead of the intellect, and you may fill in the steps that got you there only later. But in the end it has to be rational.

So the intuitive is involved in both science and religion. Do the intuitions differ in quality or only in degree?

There is a gap at least in degree—perhaps because my scientific training prepared me less for reliance on intuition. You live your religious life daily, you're constantly confronted with it, and you have been learning about it ever since you were born. But with science it's harder work, because first you have to work hard to understand the molecules intellectually.

At one time I was a chemist and worked on quite small molecules. But when I changed to biochemistry, I worked on proteins—which of course are huge. I found the transition very difficult for a long time. A small molecule is a relatively simple system; you can visualize the electrons and where they're moving. But a protein is just vast, and suddenly you have to work very hard to understand where the loops are, where the sheets are, where it fits into the membrane, how flexible it is, and so on. It's only after you've really worked at it intellectually that you don't have to keep recalling it intellectually. Then you just know it inside yourself.

Perhaps that's my ideal of science: that I will bring to science the kind of method and insights that I have in the realm of my personal religious and spiritual life. In religion there's a close relationship between my unconscious experience and my conscious thinking—so that, without too much difficulty, I can bring my unconscious experience to a conscious articulation. In science I'd like to feel that one day I could have this unconscious experience of a molecule and articulate it, give it a scientific rationale, in the same kind of way. Occasionally, in my scientific career, I suppose, I have succeeded at this task.

Isn't that how some of the greatest scientific work has been done: the scientist has a dream or a hunch, and then she formulates it in a way that can be tested?

That's right. You can't just order up such moments, and so they are very precious when they come; you treasure them because they don't happen very often. But I do think that's when I've made my greatest, most perceptive moves.

Is this similar to a religious pattern of insight?

Yes. Often the religious articulation is helped by reading other texts, as when you read a passage and you think, "This is just how I feel!" This applies not only to religious texts. Personally, some of my greatest moments when I was a teenager were with Hermann Hesse and *The Glass Bead Game*. I thought: "This is just wonderful—this is how I feel, and I've never found anyone who's said it in a way that I really understand."

It's the same principle with science: you read a scientific paper and you think, "Somebody else has thought this as well, I'm not just an odd-ball," and you discover a community that you can identify with.

So the scientific community is indispensable. Is there an equal significance to the religious community—finding other people who share the same experiences or interpretations?

Yes, and both are indispensable to me. I think religious community is important because you are surrounded by people with more experience. You have people with whom you can share your deepest thoughts without having to explain yourself all the time, because they understand the same kind of language. And, just like a religious community, the scientific community uses the same kind of language within itself.

Bringing unconscious thoughts to conscious reality is really important, because it frees your unconscious to go on and explore something else. When you write a poem or a research paper, your written work becomes a building block on which you can stand while you search for the next piece. So making the effort to articulate what we know is vitally important. Until you do that, you can't really move on, because your mind's still full of the clutter of the last unarticulated experience.

It's a bit like Theseus and Ariadne's holding on to the thread in the Minotaur's labyrinth in Crete: if you unwind the thread you may explore, but you can always find your way back to where you started from. And you can bring back with you ideas or information that you've gleaned in your meanderings. When you articulate your starting point, when you know you've proved it rationally, you can then see how the new information that you're bringing back actually fits in. Articulating is like building a secure place, a starting point, and a place to return to.

A home?

Yes, but a nomadic home, always changing. The base is never static, because every time you come back you bring something more to add to the base. I appreciate that if you approach religion the way I have, there's a danger that you could go off into some fantasy world. But if you always come back to where you believe that you've rationally established a base that you've tested with others, then I think you can afford to pursue ideas that are risky. Just as in science, the new insights that are right will come back and lock into place, and those that are junk will fall away.

I think science and religion are similar in this respect, but I think they're different too. The difference is that one's religion is exceedingly personal and individual. Religion is like a huge jigsaw puzzle: you may share all the pieces, but no two people will ever put them together in the same order. But in science, on the whole, people do fit the pieces together in a similar order. Often you must have one piece of information before you can do the next. If other scientists do the same measurements, they'll obtain the same data as I do. After a few years people will reach a consensus and will agree on an interpretation of the data. There is some sense of a jigsaw puzzle in science, but it's nowhere near as great as it is in religious life, where everybody's personal situation is so different.

In the religious life things are not as simple as in science. People may have similar life experiences, but they don't make the same observations from that life experience. They draw different conclusions, and they may well not reach a clear consensus. So I think exploring the spiritual is much more personal. Approaching molecules is in a sense much simpler, isn't it?

Earlier you mentioned testing. Could one test religious beliefs in the same manner that one might test, say, a hunch in biochemistry?

Not by physical measurement, but you could test a religious belief against life experience, couldn't you? A religious belief either works or it doesn't, although it may be a while before you discover which.

I don't accept anything in my religious faith unless it's come into focus. If people ask, for example, whether I believe in life after death, I say, "I don't know." I don't take someone else's idea from a book of dogma, then try to test it. Instead I build my faith from my own experience and spiritual life.

Take, for instance, the doctrine of the Assumption of Mary. [According to this doctrine, held within parts of the Christian tradition, Mary, the mother of Jesus, was "assumed" into the being of God at the end of her life, rather than dying a normal human death.] When I was a child I thought, I can't accept this. So I didn't include it in my own personal system of beliefs; the idea just sat there. Then I had an important experience. I grew up in a family that really wanted a boy instead of a girl. From the age of three it was obviously clear to me that it was a great advantage to be male, so I heavily repressed the feminine side of myself. When I had children myself, I recognized this in a big way. At that time I read a passage from Jung where he talked about Leonardo da Vinci's painting of the Madonna on the rocks. Here is the feminine stranded out on the rocks, abandoned and not relating to the rest of the psyche. Suddenly, I realized that this was what I had done to myself. Only then did I recover the doctrine of the Assumption , which allowed it all to make sense at a spiritual level. In the Assumption—for the first time in recent memory, or probably in the history of Christianity—you've got the feminine as part of the Godhead. No longer is it just three male persons—Father, Son, and Holy Spirit; now there's this recognizably feminine aspect of the Godhead. The idea of the Assumption allowed me to validate my own femininity. It enabled me to realize how important it is for women, and for the world, to recognize the value of femininity. So now the doctrine of the Assumption is actually important to me but only because of that experience.

That's a beautiful example. I was imagining the opposite picture: a body of doctrine received, in some sense believed, and then tested bit by bit to see which parts will survive testing. But what you've described is much more like your earlier picture: religious beliefs, accepted only insofar as rationally articulated, become solid bases on which you build other beliefs, which are selected from an undercurrent of doctrinal possibilities in light of life experience. Have I got the second picture right?

Yes, that's how I would describe my religious development. Obviously, it started as a very small child. My memories go back to when I was four years old. At that age I couldn't articulate my experiences at all, but as I got older I understood what it was I experienced. I can remember people in Sunday school trying to teach me how to pray; it never made any sense to me, because I already had this kind of inner dialogue without words. I suppose I've always had a contemplative streak.

I never learned techniques to pray. Well, that's not quite true. I did once—and it was a disaster. I was forcing myself to do it someone else's way. Then I realized that the techniques were aimed to bring me where I already had been—so I have never used techniques since, and I find that approach to religion a big turn-off.

You emphasize that you've held beliefs that have come to make sense to you, and you seem willing to say to others, "Here's what my tradition says, yet I'm really not sure about some of these claims; I'm not willing to assert them just because they're in the tradition." Would you call yourself a theist or, specifically, a Christian theist?

Yes. I believe that all of us are born with a life spark within us, which I'd like to call God, the deepest root and ground of my being. I think all of us, to some extent, display that in our inner life. But for many of us the spark is clouded because we haven't got the courage or we're not really in touch with that part of ourselves. But I think Jesus was somebody for whom there was not so much of a barrier between the inner and the outer person. He was able to show that divine spark with all its beauty and majesty in a way that very few other people, or perhaps no others, have been able to do.

How do you understand traditional beliefs such as sin, separation from God, and Christ's redemptive work as opening the way between humanity and God?

Well, I have a lot of problems with those ideas. I think you have to understand the history of sacrifice in religions—where you try to pacify an angry god or wish good luck on yourself or use a scapegoat. The idea of the god who dies for his people is very ancient and in some ways the ultimate in loving. I think Jesus came at a point in history where notions of sacrifice made sense to the people around him. However, I always have great difficulty with the idea of somebody's actually dying for my sins. I just don't want to dump the responsibility onto somebody else. I also have children. As Kierkegaard observed in *Fear and Trembling*, I don't see that sacrificing one of my children—or God's sacrificing his son—would be an honorable thing to do. I can only approach this idea through mythology.

What I do think is that the kind of life Jesus lived is inevitably going to involve making some kind of sacrifice. It's a very uncomfortable way of life to put forward to people, and mostly they don't like it. It's pretty inconvenient, and in the end—if you keep your integrity and keep to the truth—you'll probably end up dead, because the rest of the world can't handle it.

I have trouble with this question, "Are you saved?" I've always believed that God loves me as I am, and I don't have to pretend to be anything other. The motivation for not sinning is that it's so painful to be separated from the light that's within you. God, in a sense, lifts you up. The presence of God is actually what saves you, not the death of somebody.

What's important to me about Christ is the idea that humanness was taken into the Godhead, as at the Ascension, rather than that God came to humanity, as in the Incarnation. God actually raised humans up to have aspirations far beyond our wildest dreams. These longings, these aspirations to have what is so much more desirable, mean that lesser things are just a barrier if you get entangled with them. It's very difficult to know what sin is intellectually, but you do know spiritually because sin is what gets in the way.

Let me ask about the traditional belief—common to Judaism, Christianity, and Islam—that God created the world in order to carry out a purpose. The Abrahamic faiths speak of a cosmic design or purpose at the origin of the universe and of a telos, or destiny, in its future. Is this a notion of purpose that you think is compatible with the conclusions of modern biology?

Well, the idea makes sense to me: that the universe is an expression of the love of God, the mind of God, or the expressed word of God—that it's not just a blind chaos of molecules. These molecules have organized themselves into man, who is able to reflect on himself and the universe and in some sense to create his own idea of God. I think that God is at the very least as great as the sum total of mankind, isn't he?

I don't know about destiny. I would go along with St. Paul, that creation is groaning, struggling to become more perfect—whatever one means by that. In the heart of most people is a desire to reach some ideal of justice and peace, of care for each other, of leaving the dross of the worst side of human relationships behind us. It's a desire to move toward something Christians might call the kingdom of God. Whatever our religious views, we all seem to have within ourselves some concept that there's really something better than the world we now have. We try to reach some culminating point in our lives and to have something meaningful to say at the end of it. If God is at the heart of all people, then there is some personal destiny; we all try to move toward a destiny for ourselves, don't we? And if collectively we are in some sense the "mind of God," I suppose the sum total of all that adds up to a destiny for mankind.

Theologians nowadays seem to be divided between two views. The first is a traditional, roughly Augustinian view in which a perfect, timeless God stands outside the physical universe. The second is a newer, process-oriented view in which God is intimately involved in the unfolding of the universe, the evolution of life, and the development of humanity. Does your experience, as a scientist or as a person of faith, incline you to one of these views?

The second is certainly the view I would take, that God is immanent and intimately concerned with every part of humanity. One idea that's really important to me is cooperation with God. For example, I think we have the responsibility for working with God to decide whether we have children and to decide how we bring those children up. It's not like God is pulling puppet strings. However the universe evolves, it will evolve by the way nature selects itself and by the way that man interacts with nature—including how individual men and women are inspired by their understanding of God to move things in that direction.

But that's only part of the answer, isn't it? There's also the transcendent God in control of the cosmos, of the Big Bang, and of the other major events of the universe. As a biochemist I don't think I understand much about those.

If, as you suggest, God cooperates with humankind, could something similar be said about God's cooperating with the universe of physical cosmology?

Well, I would imagine yes. If God can identify with me, he can identify with a molecule, can't he? Or if I can identify with a molecule, certainly God can. But I don't have any inside information about that question, and I don't think it worries me particularly. In the end, I find it somehow difficult to believe that there's a person up there directing or preplanning it all. That belief, it seems to me, tends to take away what is the really magnificent thing about humankind: that we do have some measure of control and we do have some responsibility. If you believe that everything's predestined and it doesn't matter what you do—well, nothing will really matter.

Something that comes to mind is the book by Paul Tillich, *The Courage to Be*. Tillich grasps what is part of the greatness of the human being: that you have the courage to have dreams, even though you know your personal failings and you know that you're going to die. We know that one day the sun or even the universe will no longer be here, so what we do won't really matter. Yet we have the courage to go on. That we have this courage is an indication of the greatness of the created world, isn't it—that we still believe that it's worth being just and honorable.

It seems as if you don't find any incompatibility between the physicist's account of the process of cosmic evolution and the religious account according to which God was directing that process.

No, I don't, because how would we know? All I am sure about is that I experience a God who cares, who really interacts with me personally and with the world. If people are killed in a natural disaster, he is there, suffering with them. One of my children is in a wheelchair, and when that happened, I was convinced that God was suffering with us. There's a natural world that has its own laws. Maybe God intervenes, but I don't really understand why he should create a world and then interfere in the natural creation, as though he had somehow got it wrong in the first place. I have trouble with that.

If you ask me what happens when I die, I will say, "I don't know." But I do know within myself that the God who has loved and cared and interacted with me—since before I was old enough to put a name to it—will continue to do that after I die. I don't really need to know any more than that. If as scientists we learn more about the universe and we begin to see a cosmic destiny, that would be very interesting. But I don't have any preconceived ideas that would be threatened if science found it one way or the other.

Christianity, like Judaism and Islam, has emphasized that humans are beings with a special freedom and moral responsibility and capacity for relationship with God. I wonder how you see this view of what a person is as fitting in with the biological study of organisms?

Biology doesn't even attempt to address that, does it? I think most people would feel that they are individuals, and they do have responsibility, and they do have free choices. I'm sure that if you moved from biology into the social sciences, you would begin to talk about what it means to be a human being. But just as chemistry doesn't really talk about biology, biology doesn't really talk about social sciences.

If a biological process is very simple, there's generally only one way of doing it. For example, in mammals the taking up of oxygen by hemoglobin in the red blood cells is quite simple and mechanistic. On the other hand, a human being looking for food has a hundred different ways to do it. What you choose to eat at any one meal is very complex. The more complicated the organism, it appears, the more choices to fulfill the functions and the needs of that organism are present. At that level, choice or freedom to select between things is essential to survival.

I would, however, like to confront the idea that people are just doing what their genes tell them, that they're simply defined by their genetic inheritance and their environmental makeup. I think people are actually more complicated than that. People make many different selections between many options every day; they choose, and not necessarily in totally predictable ways. I can't believe that any people in their personal life actually feel that they don't have any kind of free choice—even though their choices may not always be conscious but rather more intuitive.

It seems to me that organisms need to have free choice, and they need to be adaptable. Human beings share this need to adapt to new environments with all the other higher animals. One way of structuring your community may be fine when the weather's dry, but if you suddenly get flooded out in Bangladesh, then you have to reorganize and regroup very quickly. In the human case it's got to be a matter of conscious choice, because merely reacting unconsciously to your genes certainly wouldn't give you the same flexibility. So I don't think genetic programming alone makes sense.

Your argument against genetic determinism doesn't seem to rely on ideas imported from outside biology. Can the same kind of strictly biological argument be made for moral responsibility? Is there anything you could say, as a biologist, about the sources of moral responsibility?

Well, I think some moral responsibility is biologically implanted. But then there's altruism, isn't there? There are many examples of people who have put other people's safety before their own, which conflicts with the survival of the individual. I suppose you could say, "They're just doing it so the species can survive, even if the individual does not." A biologist might even say you need your child to survive to propagate your genes, so you will be compelled to put your child before yourself.

But we're really talking about moral choices, aren't we? Look at

Abraham's commitment to offer God the best that he had, leading him to consider sacrificing Isaac and his hope of descendants. People do some very irrational things that don't immediately help the survival of the species, so clearly there must be other mechanisms than just survival. People have beliefs about what constitutes survival, and those beliefs might not coincide with biological reality. Maximizing survival may explain a great deal, but I don't think it's adequate to explain the whole of human behavior.

I was just reading Primo Levi's terrible accounts of the concentration camps. Even there, people gave food to others although they were depriving themselves. Some felt guilty afterward for surviving when others didn't. None of that fits with wanting to survive at all costs and not caring about anybody else, does it?

Why would human beings spend time painting? Why do we live beyond the years of reproduction? These aren't essential to survival—or you have to tell a lot of roundabout stories to explain why they are. To convince me that we are only responding to biological stimuli, biology would have to explain the purpose of the species in rather bold detail.

Many people hold notions about what humans are that transcend anything that biology can tell us, and for some their notions are clearly derived from their religious beliefs. Is that an approach that you would follow in attempting to understand the human being?

Apparently, humans are the only species that can reflect upon itself, and that seems to make us different. We also have much greater powers to articulate and to communicate with each other about ideas; it seems unlikely that other species can communicate abstract ideas to any meaningful extent.

The mind has to represent more than the sum total of the biology of the person. Mind belongs to the person, but it is also *beyond* the person and in a sense *something other than* the person. For instance, the mind isn't consciously aware of all the chemical processes that are going on in the body—it's beyond them. Its interactions with the body, the community, and the world are so hugely complex that it's hard to imagine that the mind could simply be programmed. And it's hard to believe that the mind could merely reflect a blueprint, because every person is so different from every other person.

Maybe I think that something deeper inspires the firing of the synapses. I don't want to use the word *soul,* in the sense of locating a soul in some particular part of the human being. But I feel that a human is more than the sum total of its biological parts, just as I feel that God is far more than the expressions of God in us.

Human community is like that too—more than the sum total of the individuals. A crowd of people has a life of its own, its own ways of responding to situations, in which the individual selves are subsumed. Playing a musical instrument is like that too—the tune is beyond the sum total of the notes you play, and something has meaning beyond the notes. I suppose I think a human being is like that.

Some philosophers describe these properties, where the whole seems greater than the sum of the parts, as "emergent properties." They speak of life as an emergent property of complicated molecular structures. They speak of mental properties as emergent properties of the brain. When we talk about the spiritual or the religious, could these be further properties that emerge out of lower levels?

I think the spiritual or the religious does evolve, does emerge from the lower levels of biology. If you quote the traditional doctrine, Jesus is wholly man and wholly God; I think of that as the emergence of man into the Godhead. The ability to experience a spiritual dimension grows out of being human. I see it that way, rather than the soul as being something that dictates to your biological organism what you do.

You said that the spiritual evolves. Would this process of spiritual evolution be at the individual level or at the species level or both?

Have you read a book called *The Myth of the Goddess*? It is quite speculative, but according to the book, the perception that there was more to life was originally understood by a relatively small number of people who had a particularly strong interaction between their conscious mind and their unconscious. The early shamans managed to articulate those dreams in a way that was meaningful to their people. They developed hunting rituals or fertility rites or vision quests. They told stories and painted pictures in imposing caves. They helped people to have experiences so dramatic that they were imprinted on their memories for life. From there this spiritual side grew; people found that it helped to have a mythology to which they could relate their life experiences.

That's part of what religion is for me: a mythology I can relate to, so that I can feel that I'm part of something that has happened before. In a very lonely universe I can find places where I've got a foothold. At a lower level of survival it can be in response to dire straits. But at the greatest moments in my life, it's a positive experience that grows when life is good.

Let me ask explicitly about divine action. The theistic traditions have held that God is active in the world. Many thinkers, including some theologians, respond that our increased scientific knowledge of the world leaves little room for any sort of direct divine action. Do you think belief in divine action in the world has been supported or challenged by the growth of science?

In the past, people thought that God directed the floods of the Nile or caused pestilence. We don't think that way anymore. But each generation has its own needs and interprets religion for its own time. I'm very aware of the presence of God—not directing everything but being there to meet me when I need to grow. God's presence is like a friendship: something very strong requiring the best from me. This describes it better than the idea of a God who comes in miraculously to solve my problems.

So it is in these key moments of human development, in which you find God, that we can speak of divine action in the form of God's presence?

Yes. Sometimes you get to a point in life where the way you're living is no longer appropriate because you've matured or circumstances around you have changed. For instance, when the children leave home and suddenly you have another part of life in front of you, you need to reassess and rethink. You need to fall apart and reassemble. At those moments I find God. I have a lot of really helpful dreams—when I am searching for some guidance, some signpost—and that's where I meet God.

Is it misconceived, then, to think of God's acting to alter natural events or processes?

Clearly, God doesn't work unnatural miracles, although sometimes you may interpret them that way. No, for me the divine role is that whatever happens, God will support you to handle it. It's that you can look to God to give you the courage you need in order to cope with whatever life turns up.

Even if some of the major precepts of Christianity were found to be untrue, it still wouldn't affect my faith. It's the truth of what is conveyed in the life of Jesus, not in the precepts or assumptions, where the meeting place is for me. At the same time my religion is a bit eclectic. I find a lot of things in Hermann Hesse very relevant, yet in his novel *Siddhartha* the focus is on Buddhism. When I went to Japan, the Silver Temple and Zen garden were amazing and almost overwhelming—like walking in a three-dimensional painting where I knew I had always been.

My religious experience is strongest at moments of inspiration, when I am trying to reach the unknowable—both in science and in myself. It's like the story of Parsifal and the salmon. He tasted the salmon and then dropped it because it was too hot to handle. He couldn't interpret it, but for the whole of his life he searched for that experience again because it was so precious. I think that's where I am with religion. Once you've been touched, it's so desirable that you'll sacrifice almost anything else to experience and understand more.

Tell me more, if you would, about those moments of inspiration. Do you derive religious inspiration from your scientific work? Or scientific inspiration from your religious practice? Or both or neither?

Does science contribute to my religious thinking? And does my religion contribute to the way I do science? When I write creatively about my spiritual development, I often use scientific metaphors and analogies. For example, I write about my relationship with God's being like DNA, twisted but then separating. But the contemplative experience is more powerful than visual metaphors. When I'm finally in tune with a molecule, I no longer have a visual picture of it—I have an experiential picture. When I experience God as truth within myself, I don't have a visual picture of God—it's an experiential picture. Sometimes, when I'm drawn into a religious contemplative state, I am also drawn into a scientific contemplative state that is very similar.

Do you find that your contemplative practice on the one side, either religious or scientific, leads to greater depth of understanding on the other?

Yes, although I think that the contemplative state is not something I've learned so much as something that I've always known, my natural way of being. The great religions have always known that several things are crucial to the contemplative state: austerity, discipline, and humility. At the deepest level I think those same things apply to science.

By austerity I mean that you have to pare away the nonessentials, so that you're looking at a tree without leaves in winter—just the bare essentials, no pretense. Not trying to imagine that it's what you'd like it to be but rather just being with what really is. This is crucial in spiritual development, and it's also crucial in science. In that sense I take what I've understood from my religious tradition into science and certainly the other way around.

At the end of my university studies I began a very close association with a convent. I was attracted to it because I sensed these were people who understood how I was thinking. They were much more experienced than I was; they could affirm what was emerging in me and help me to grow in it. Today this community remains a strong part of my life.

I did consider that I might have a religious vocation. At the time I was going out with an ordinand who then became a priest; we were very much in love and wanted to have children. On the other hand, when I was about thirteen I had a powerful experience that committed me to science for the rest of my days. By age twenty, then, I had three different strands to my life: a religious calling, a scientific calling, and a calling to marriage and parenting. I went to the Mother Superior and she said, "If you want to become a religious, you'll have to give up everything else." I went to my professor and he said, "You know science is all consuming." I went to my mother and she said, "If you want to have children, you'll have to give up everything else." This was in the 1960s and in those days, if you were a woman and going to have a career in science, you had to consider seriously whether you should have a family, and you certainly couldn't be a religious.

I was left with a big problem: I knew that whatever I decided, two-thirds of me was going to be unfulfilled. So I went to the convent and became an associate. Remarkably, they accepted this threefold commitment. My whole life, in short, has involved living out these three callings. It's like a plait of hair, with three different strands coming together. Sometimes I may go deeper in one part—the science, for example, may develop at the expense of the others—but then I reach a point at which

I can't go any further with science unless I understand some other facet of myself.

I suppose I am quite an intense person; life for me is quite serious. There are many things that I feel are important to pursue, and I can't always do them all at once. But in the end I can't live without all three strands. They've never been in conflict with each other, but sometimes my lack of maturity in one has inhibited the development of another.

You said that the development of reason on the scientific side has allowed you to think about religion in a richer way. You also said that the development of intuition or contemplation has contributed to your scientific discoveries. You seem to have developed a positive cross-fertilization between the three areas. They almost seem to work together organically.

Yes, yes. And the possibility of integrating them stems from an experience that I can only define as love. The response that I had to science when I was thirteen was purely one of the heart, just as I fell in love and just as I responded to God. Although the intellect follows quite quickly and analyzes it, the initial response is from the heart. I would also like to say that it wasn't that I sought God but that he sought me; and it wasn't that I sought chemistry but that chemistry sought me. When I came in contact with them, something inside me leapt and recognized that that's where I wanted to be, where I belonged.

9. Satoto

Dr. Satoto is an expert in nutrition and child development from Indonesia whose work has shaped public policy and improved the lives of children in several countries. First trained as a physician, then as a clinical nutritionist, Satoto (his full name, by Indonesian tradition) did his doctoral work on the connections between nutrition and child development. Raised in a devout Muslim household, he considers his research a path of devotion.

Satoto directs the Research Institute at Diponegoro University in Semarang, Indonesia. The institute, on the island of Java, comprises eight research centers that examine environment, population, energy and natural resources, technological development, development

Philip Clayton interviewed Satoto; Jim Schaal edited the interview for publication.

studies, sociocultural studies, health, and gender. He also coordinates the Indonesia Partnership for Child Development, which began a long-term survey of the health, education, physical fitness, and cognitive development of fifty-five thousand children in 1995. The program has since developed a school health insurance program for thirty-four districts of Central Java. Satoto is chair of the Indonesian Epidemiology Network, a group of institutions committed to finding solutions to population-based health problems through the application of epidemiological, social science, and health management approaches. He is also secretary of the Center of Excellence for Control of Iodine Deficiency Disorder in Indonesia. During a distinguished career as a scientist and humanitarian, Satoto has worked with UNICEF, Helen Keller International, the World Bank, and a variety of government and nonprofit agencies in Indonesia, West Timor, and Bangladesh.

Even at first meeting, Satoto exudes the aura of a physician. One senses an inner calm and focus; if there were an emergency, this is the kind of person you would want to come upon the scene. In contrast to the carefully reflected, cautious comments of some of the other scientists, Satoto's responses are immediate and disarmingly honest. Where he has doubts or sees inconsistencies, he states them without embarrassment, perhaps reflecting a faith deep enough that it is not troubled by this or that particular intellectual conundrum. It is humbling to sit across the table from a person whose integration of Islamic principles with good nutrition practices has single-handedly saved millions of children from malnutrition.

How did your background in medicine and nutrition lead to your work in child development?

After medical school and a degree in clinical nutrition, I did my Ph.D. thesis on the relationship between nutrition and child development. At the time I was still very young, and I was not satisfied to understand nutrition only, so I entered a new world with child psychology. Finally I was able to make the connection between the two fields in my research on the influence of nutritional adequacy on child development. When one says child development, one means not only physical development, as in growth and motor development, but also social, emotional, and of course intellectual development—all taken together as a single system.

We found that nutrition affects not only the survival of children but also their development too. For human beings survival alone is not enough, right? Just to have a child alive is good, but what if the child cannot learn or perform in school?

For a child to learn well, it doesn't merely need enough calories in the daily diet, does it? Doesn't it also need specific nutrients?

Both. Good nutrition requires more than the adequacy of energy in the form of calories. Originally, we concentrated on energy intake and ignored micronutrients, but studies from the late eighties up until now show that energy alone is not very important. Calories have a general influence on child development, but some specific nutrients—protein, vitamins, and minerals— also have an influence. Take zinc, for instance. In the development of brain cells, many enzymes are involved, and enzymes are composed of micronutrients such as zinc, selenium, or vitamin A. To take another example, there was a fairly big deficit in vitamin A and iron in a certain group of children that we studied. This is very serious: that generation has lost not only intellectual ability but also the ability to live normally. Energy alone is not enough; a balanced diet is necessary.

I worked together with Helen Keller International to do a nutrition surveillance, examining a group of children and mothers once every three months to measure the adequacy of vitamin A, zinc, iron, and things like that. By having a voluminous number of samples, we can reach some conclusions about the relationship between micronutrients and child development.

As I understand it, you were able to bring the results of your research to the attention of the government in order to improve children's diets.

Yes. That was a triumph.

I would be interested to hear about your work on the effect of religious beliefs on diet and family planning. Could you address that issue?

By chance I had some experiences in dealing with the relationship between nutrition and family planning in Islamic communities. Many years ago family planning was introduced to Indonesia, and it failed. Some anthropological research was done to see why it didn't work out, and we undertook some efforts to educate people in Indonesia with regard to family planning.

Approximately 75 to 80 percent of the population in Indonesia are Muslim, so we tried to find a passage in the Qur'an relating to family planning. We worked together—the religious leaders and the scientists—trying to make people understand that family planning is not killing children or fetuses in the womb but is a long-term effort to create happy and prosperous families. We kept trying over and over again, first in a pilot program and then nationwide. In the history of family-planning programs, it was really one of the best achievements. One reason is that we used religious motivations to encourage people to say, "Family planning is good for everybody, because it is in line with our belief in Allah."

Now we're trying to use the same approach for nutrition. If we say, "This food is not good, this is low in calories," people just don't listen to us. What do they care about calories? They care whether a food is delicious or not. They care that the food should be acceptable to God *(halal)*, and not be prohibited (not *haram*). We were working with the same institution, the Family Planning National Board, which also has a huge nutrition program involving approximately ten million children all over the country. Through the program we introduced good nutrition from an Islamic viewpoint.

That is fascinating. When you addressed people's religious motivations first, you had dramatic success in teaching them good nutrition and family planning. In the West, especially, scientists might not concern themselves with such religious motivations.

Of course, child development is not limited to physical or intellectual development. We are talking also about religious or moral development, which is often neglected among all the other types of development. In Western society experts dealing with moral development actually start from the angle of psychological theory. But in Asia, as I experienced for many years, we are not starting from the theory but from what we call the practice of religious living. These are two different horizons.

In Asia we don't teach people psychology or the theory of moral development. We deal with religious perceptions, with religious realities, with the lifestyles of the people. All educational texts on nutrition are related to religious messages. We say to people, "Come on, eat the best nutritious food, so you will live happily in the world and then you will live in the afterlife." I am not an expert in religion, but I believe this way is correct in relation to what I believe as a Muslim.

Have you found that this approach—integrating religious motivations with scientific nutrition and child development theory—can be used in other Muslim nations?

Hypothetically, yes. I was hired by the World Bank as a consultant in child development for Bangladesh. Bangladesh is a Muslim country—at that time it was more Muslim than Indonesia—so I brought in Islamic beliefs in order to motivate people. But actually the process became rather difficult, and I was somewhat disappointed in the end. Most people were not familiar with the integrated concept of appropriate science and religion. My Bangladeshi partner explained to me, "It is very difficult because you are the first person who is trying to use Islamic belief to explain something physical, rather than something that is strictly religious."

I had another experience in West Timor. Most people in West Timor are Catholics, so I worked with a priest there and tried to use phrases from the Bible instead. This worked better than in my experience in Bangladesh. Because I don't know the Bible well, the difficult thing was to find a good example or analogy from the life of Jesus to use as a motivational tool. But in general the approach of using Bible quotations was well accepted. The first point is to find something from the Qur'an or the Bible, or whatever—in a Buddhist country it might be the Tripitaka. The second point is to understand fairly clearly the religious style of a particular people. There are different interpretations of religion between one community and another, even when the religion is the same.

Are there other areas where Muslim beliefs can be integrated with community development or social programs?

Although it didn't work in Bangladesh, using the principles of Islam may work in some places—if the application, like nutrition, is not dangerous or embarrassing. For instance, there is a kind of Muslim bank, based on rules taken from the Qur'an, called a *sharia* bank. There are more and more *sharia* banks in Indonesia, especially the small retail banks.

But sometimes we are suspicious. Take something like nutrition: Is it really a clear, honest integration between religion and science, or is it a way for the scientist or bureaucrat to motivate people to do something by using religion? It's very difficult to answer this question, but sometimes I ask myself: Am I honest when I integrate my belief in my religion and my daily scientific work?

That's a very good question. Are there some areas where you find it easy to connect your own scientific training and practice with Islam and other areas where it's difficult to link them together?

Let me put it this way: I have been a professor for twenty-six years, working in government, and do you know how much I am paid? Less than two hundred dollars a year. So for me science is a spiritual interest.

I believe in God very much. If God is real, if God exists, then that's one of life's truths, right? Then all truth comes from him. Three or four years ago I realized that God gives us so many ways to find truth. Religion is one of the ways, but scientific truth is another way. When I give lectures, I always say, "Truth comes from God, and what we are going to talk about now is actually small drops of the truth of God." There are still many things we don't understand, like cancer, and sometimes in our lives we find "truths" that do not in the end count as true. Yet God gives us the continuing opportunity to understand.

Can you tell us about your own experience as a scientist and a Muslim? Is it easy for you to integrate the religious and the scientific sides of your identity?

I was born in a Muslim family. Since I was very young, my father and my mother taught me how to read the Qur'an, to understand the ceremonial aspect of Islam, and to understand why God asks me to do this and not to do that. Over time it has become truth in my life. It would be very difficult for me to change—to become Catholic or Hindu or something else—because for many years there was the indoctrination that truth is what God gives to Muhammad.

Sometimes it seems as though the truths that one has been taught are beyond question or at least they are very difficult to question. When I was in high school, I had thousand of questions. My German language teacher at the time told me, “If you use your brain to understand the truth coming from your religion, you will fail. The issue is, do you believe or do you not? If you do believe, that’s all that matters.” He said if I did not believe, then I should try another way to find another truth, another source.

Later, when I became a doctor, I began to understand more. At that time I was being supported by UNICEF to study the relationship between religion and nutrition in Islam. I worked very closely with Gus Dur, the president of Indonesia, in his private library when he was at the religious school. We found many things in the Qur’an that are correct from a nutritional point of view—things like “You can eat, you can drink as much as you can; but do it just adequately, do it just enough, because God does not like people who are excessive.” Here my understanding of nutrition was in line with my religion: if you eat more, God doesn’t like that, and you become obese; if you eat less, God doesn’t like that, and you get malnourished. Of course, there are some places where contradictions seem to arise, and there are many interpretations. In my experience, we can interpret contextually, in a way, so that nutrition science comes very close to the truth of what God told Muhammad about his daily food.

So as you advanced in your scientific training and in your study of the Islamic tradition, you found that the two supported each other?

Yes, exactly. I feel I’m on the right track, that I’m doing right in front of God and in front of the community in my scientific discipline. When the people said, “Any good thing that happens to you comes from Allah, but any bad thing that happens to you comes from yourself,” Muhammad replied that they should say instead, “Everything comes from God, and to God it all returns.” But [*laughing*] I heard from a British doctor that “God cures, and the doctor takes the fee!”

Yet one might well wonder whether the contribution of science and the contribution of religion are both on the same level. Is the explanation scientific and the underlying reason religious, or does the Islamic tradition offer something equivalent to empirical science?

Science gives us more detail. But any time I open the door of understanding with a lamp, I believe it is only God that can give me new insights. God gives these through what Islam calls *sunnatullah,* the natural laws of life and the rules for finding them. In scientific terms God gives insight when we follow the correct methodology. If we do experiments to eradicate malaria, for instance, then God will show us what the answer is if we follow the *sunnatullah.*

But even if another person does not believe in God, yet he follows the correct scientific method, then he will find the same thing. This is where science cannot be compared to other ways of finding truth. And this is why ethics matters in science: anyone who knows the right methods can make atomic bombs. The truth depends very much on the objective of the research, right?

But aren't you saying that something is gained when one uses the right scientific practice and at the same time does his research in the name of God?

Yes. Compare this with what a humanist does. A humanist keeps a high value on humanity, and when he does something that he believes is dedicated to humanity, he will be happy. The same thing happens to the believer from any religion: when you do something and you believe it will please God, you will be happy. When I am doing medicine or nutrition, when I say to God, "I am doing this and I hope you are pleased," then I am happy in the world. Besides being paid well and being popular, I will be blessed by God. This is an added value of asking God to be involved in the whole scientific process.

That's nicely put. Let me now ask you a somewhat different question. The scientific attitude is rational, critical, methodical, and some would say objective, but the practice of religion seems very different. As you went through medical school and scientific training, did you struggle with this difference?

In the university in Indonesia the emphasis was on rationality, and religion was not worthy of mention at all. In medical education the curriculum does not involve God at all. But some professors at least try to give added value by telling the students that in practicing medicine they are doing good works. I'm happy in medicine, in health and nutrition, so for me there is no contradiction. Maybe in physics or other scientific disciplines you can find more contradictions, but in human sciences, such as medicine, health, and nutrition, we don't actually experience such contradictions.

For example, Islam does not allow the believer to eat pork. So some doctors are trying to explain why we are not allowed to eat pork. Looking only at nutrition doesn't work, though. Another food we are not allowed to eat is blood, but blood is very rich in iron, vitamin B_{12}, and in protein. So we have to find another rationalization, if you wish to call it that. From a religious perspective, blood might reflect cruelty, which God doesn't like. Well, according to my interpretation, pork is no less healthy than beef or chicken. Rather, the rule against it is symbolic. By asking people not to eat pork, the idea is not to behave like a pig.

So the religious practice is not about science. It's about ethics, about how you live.

Sometimes yes, sometimes no. Fasting is the newest thing. It is very difficult to say that fasting is healthy, because from a nutritional point of view it is not. Not eating or drinking is not healthy at all, right? But then we find that fasting trains our nervous system to control our hunger and thirst. If we can learn to control these drives when we are safe, then when we find ourselves in the situation where there is nothing to eat or to drink, we can remain safe for some time. In this case a moral interpretation works better than a nutritional interpretation.

Scientific truth should be challenged, because scientific truth is correct until somebody else later proves that it is incorrect. In the era of Cleopatra, when somebody had diabetes, they were told to eat a lot of sugar. It was thought that you lost sugar because it went directly through the urine. That was the scientific truth at the time. Then we learned that in diabetes the body doesn't have enough regulation to control sugar in the blood, so some of the sugar is lost through the urine. So now the treatment is to restrict sugar, right? No, because now we know that the problem is not just sugar intake but metabolism; the treatment is not to restrict one kind of sugar but to restrict calories.

So far you've found ways to reconcile nutritional science with Islam. Are there other areas of science that are more difficult to accept as a Muslim? What about Darwinian evolution, for example?

Actually, there are some ways in my religion to deal with this. Take Adam and Eve, for example. It doesn't look like Adam was one particular man or that from the rib of Adam. God created one particular woman named Eve. The question is whether the story refers to a particular physical event or whether it is primarily symbolic. In my personal belief, Adam represents a phase in which evolution arrived at the perfect creatures, namely, human beings. In the Bible or the Qur'an, God sent Adam and Eve to Earth because they ate the fruit that he didn't want them to eat. But perhaps this was the time when the Creator said, "You are ready to be on the earth and responsible to me."

Is there a tension between genetic engineering in science and the view of the human person in Islam?

In my own perception, there is no contradiction. I am very liberal. I say, why not? We once thought that cretinism was due to an iodine deficit in the diet, but in one village we found many babies that looked like cretins and determined it was not due to diet, because their diet was not deficient. My comment was that God is doing genetic engineering in the village. If God is doing that, then we can do the same thing. The question is: Is it for good or for bad?

Finally, are there ways in which your religious practice and belief help you as a scientist, or are there ways in which your scientific practice contributes to your religious practice and belief?

Yes, both. From the beginning I have been doing both science and religion together. The more I learn about nutrition, the more I believe in God, and the more I believe in God the more I am trying to understanding "the soul" of nutrition. I do my prayer while I do my nutrition research. I do my nutrition as my scientific task and also to please God.

What looks very dogmatic in religion is usually the ritual practices—going to mosque every Friday afternoon, saying set sentences in prayer, and so on. To the scientific way of thinking, it is very difficult to understand or to explain the meaning of these ceremonies. As scientists we may even ask, do we need these kinds of ceremonies? There are some outstanding problems between science and religion, and I have a long list of them, actually. But we need more time to bring them closer, to make them into one truth.

10. Paula Tallal

Dr. Paula Tallal is a noted experimental psychologist as well as a practicing clinical psychotherapist. Her primary theoretical work, on auditory processing dysfunctions in children with language-development impairments, offers a new model of the relationship between lower-level sensory functions like hearing and higher-level cognitive capacities like language. Because this model demonstrates that some language impairments, such as aphasia and dysphasia, result from relatively basic neurological problems, it challenges the widespread notions that "speech is special" and that this ability can be traced to an innate "language module" in the brain. Equally important, Tallal's research has led to valuable therapeutic applications, including

Philip Clayton interviewed Paula Tallal; Jim Schaal edited the interview for publication.

remedial language-learning software that is yielding impressive results for thousands of children.

Born in Texas, Tallal was eighteen when she moved to New York City, where she attended New York University and majored in art history. A summer job working with adult aphasics, patients who had had a stroke and could not use or comprehend words, sparked her interest in language and the brain.

After moving to England and working in a research laboratory for a time, she began her doctoral studies in experimental psychology at Cambridge University. Just three years later she had completed her dissertation on central auditory-processing disorders in children with developmental dysphasia. During this period she added Christian practice to her Jewish heritage.

"I incorporated Christianity into my personal religious beliefs when I was in graduate school," she recalled. "It seemed a natural progression, just as the Old Testament preceded the New, and Christ was born Jewish but 'died a Christian.' What I found most compelling then, as well as now, is having a personal relationship with Jesus through prayer—the word became flesh."

She spent the first ten years after graduate school in basic research on brain development in children and taught experimental psychology at the University of California–San Diego. During those years Tallal also enrolled as a student there and became a certified clinical psychotherapist.

"I have found working with individual patients as well as with families to be extremely fulfilling. It fills a very different spiritual need than research does," she said. "It also allows me to use my gifts in the direct service of another. Unlike most basic research, clinical practice is immediately rewarding because you share directly in the life explorations of others."

Now at Rutgers University in Newark, New Jersey, Tallal began working with Dr. Michael Merzenich, best known for his research on "brain plasticity," the human brain's remarkable ability to change its own structure in response to learning and adaptation. Together they applied Tallal's model of auditory processing and Merzenich's research on brain plasticity to help children with developmental language disorders. Specifically, they developed software that would slow rapid consonants and other speech sounds, allowing these children to hear and thus understand words clearly for the first time. Tested in controlled studies and replicated in clinical settings, Tallal and Merzenich's techniques proved enormously successful. Responding to growing demand for the software from therapists and parents, Tallal and Merzenich cofounded the Scientific Learning Corporation to develop and market the reading and language-learning program Fast ForWord—for which Tallal was recently awarded the Thomas Alva Edison Patent Prize.

While running the company, Tallal remains an active professor and researcher at Rutgers, a leader in professional societies for psychology and neuroscience, and an influential voice in nonprofit and governmental agencies serving children with developmental language disorders and learning disabilities. These commitments effectively join her scientific interests and experimental practice with her humanitarian motivations and clinical practice. Tallal's work gives children their voice, and in her work she hears a calling.

Let's begin with your earliest interests in science.

I got interested in science because my aunt, Dr. Lisa Tallal, invited me to stay with her in New York City for the summer after my first year of college. She got me a job in the biochemistry research laboratory of a good friend of the family. What I learned that summer was that I really wasn't cut out for biochemistry at all. It took me the whole summer to figure out that pipettes came in different sizes. I messed up quite a few experiments in George Acs's lab, and we laugh about it to this day, because they were quite certain that I wasn't cut out to be a scientist.

Were you a science major at the time?

Well, initially, I was a psychology major in my first year at a small teacher's college in Texas, where I grew up. But after that first summer in New York, I wanted to carry on living with my aunt, so I transferred to New York University. At NYU I actually majored in art history, but I worked in research laboratories to fund my education.

Of course, I hear a lot of art history majors do that.

Sure. My aunt lived close to Rockefeller University, so I was able to get a job in Dr. Neil Miller's lab there. He ran a fantastic lab. It started off as a weekend job, coming in Saturdays and Sundays so that all the scientists didn't have to come in. I took animals out of their cages and ran a variety of different experiments; most of them were conditioning studies, trying to work out the neural circuitry for eating and drinking behaviors. Then, over time, I began to work there about half time. I had a lot of opportunities in that laboratory.

It wasn't until over twenty years later that I had a chance to talk with Neil Miller, when I gave the presidential lecture at the Society for Neuroscience. Afterward, Neil came over to say how much he liked my lecture. It was really fun to inform him that I was that young art history major whom he had allowed to work in his lab so many years ago and to tell him how much that opportunity meant to me.

In my second summer my aunt Lisa got me a job on the aphasia ward at King's County Hospital. I worked with adult aphasics, patients who had lost their language ability due to stroke. Because I love talking, the idea that you could actually lose your ability to communicate was really horrifying to me. But in retrospect I think that kindled my interest in language and the brain.

That interest in language and the brain has been central to your research. Tell us about the evolution of your own research program.

In science, whom you happen to get as a mentor can be very important. That certainly was true in my case. While I was at Rockefeller, I was fortunate enough to meet Dr. Geoffrey Grey, who at the time was chairman of the department of psychology at Oxford and was on sabbatical at Rockefeller. He convinced me that I should go into science, that I was good enough at research to do that. He also convinced me that I would really thrive in the English system, which involved less direct coursework and more independent research. He must have sent out some letters about me to colleagues, because I ended up with a job at Cambridge University as a research assistant with Allan Findley in the department of anatomy. I learned a lot of basic techniques there and really enjoyed the physiological part of the animal studies. We studied nursing behavior in rabbits, and my first scientific publication was on the timing mechanism that controls nursing behavior in rabbits. Each rabbit nursed for precisely the same amount of time each day; if a rabbit nursed for two minutes and thirty-seven seconds one day, that's about what they did every day. We tried to figure out what was actually controlling that time clock. Throughout my career my research has focused on two areas—language in the brain and timing in the brain, and the neural circuitry underlying them—and these studies started my interest in timing.

But rabbits nurse for only three minutes a day, so I did have a lot of free time on my hands. As a result of this work all my friends had bunnies by the end of the experiment, because I couldn't stand to kill these bunnies afterward. Then we did a bunch of experiments on how you train a bunny not to go to the bathroom on the floor.

Those experiments being a bit more urgent in your friends' homes.

Yes, we did operant conditioning at home!

Yet you wound up studying children, not rabbits.

Since I had a lot of spare time, I started taking some classes at Cambridge. I took Dr. Mark Haggard's class on language in the psychology department and found that I was really interested in it. During that same period I was trying to decide whether to go to medical school or to graduate school. My lack of ability in chemistry pretty much determined that I wasn't going to medical school. The year that I was applying to medical school was the year in which deferments for Vietnam were changed from graduate school to medical school only; that made it even more difficult than usual for a woman to be accepted to medical school.

Based on my experiences as a research assistant and the courses at Cambridge, I decided that I really did like experimental psychology and that I would like to try to get into the graduate program at Cambridge University. In the United States there was a very formal admissions process, but in England it was pretty informal. I went to see Dr. Oliver Zangwill, the chairman of the department of experimental psychology, to find out what one had to do to apply to the graduate school. I was ushered right into his office, and Dr. Zangwill asked me, then and there, what I wanted to do for my thesis. I didn't have a clue! I said I wasn't sure, and he said, "Well, I'd need to know before we could make a decision, so come back next week."

So I rushed to the library and started reading journals. I knew I was interested in language, and fortuitously I came across an article by Arthur Benton about children who had a developmental form of aphasia. I thought that was both amazing and horrifying: it's one thing to lose your language when you are an adult, but not to develop language would have a profound effect on every aspect of a child's life. I thought this certainly would be something worth studying. Benton hypothesized that the language problem might result from problems in some form of higher central auditory processing. I decided I would like to study central auditory processing in children with aphasia, a topic that has formed the basis of my research throughout my career.

I went back to see Oliver Zangwill. Clearly, I wasn't all that well versed in the field, as I didn't realize that he was one of the leading figures in the field of aphasia. I start talking to him about this incredible disorder called aphasia; he sat there politely nodding. I lived to be very embarrassed in retrospect about this. But I also told him about a childhood form of developmental aphasia, which was newly described at that point, and told him I'd like to study the auditory processing components of this congenital language disorder. I was accepted as a graduate student and, three years later, got my Ph.D. with my thesis on central auditory-processing disorders in children with developmental dysphasia.

Since developmental dysphasia was just gaining attention, how did you approach your research?

Initially I thought I'd be able to go to the research literature, find a model of central auditory processing, and apply that to studies of children who were having difficulty learning to talk. But it turned out there really wasn't much of a model. Some studies had been done in animals, primarily with primates and cats. But far less was known about audition than about vision, for example. Of course, I went to the literature on human subjects as well. But I found that although there was certainly quite a lot of psychophysical data on how humans hear things, it hadn't really been connected to the physiology or the neurology. Nonetheless, by piecing together what was known from animals and humans, I developed a hierarchical model of how the auditory system might work.

You developed a hierarchical model of central auditory processing that didn't exist in the literature at the time?

Right. I began with the simplest aspects of sensation—detecting that a signal occurs—and moved step by step into more complex aspects. I pulled together information that did already exist, but across several different fields, and filled in missing parts.

But you ended up having to do much broader work in neural modeling than you would have thought was necessary for the aphasia study.

Absolutely.

The application of your basic research to the problem of developmental aphasia is an interesting example of the interplay between the theoretical and the applied in science. Did you intend for your work to have such important medical applications?

No, although I knew I did want to work with humans, especially children. Originally, I was just trying to come up with a model of central auditory processing so that I could test Benton's original theory: that dysphasic children had language problems because they weren't really processing the acoustic signals that are necessary for speech. I wanted to figure out which aspects of acoustic processing might be deficient and exactly how this deficiency might affect language development.

What I found out was that the processing constraints had to do with how rapidly the stimuli changed in time. I was able to show that children with developmental dysphasia were quite normal in responding to acoustic signals if they were presented with signals of long-enough duration with long-enough intervals between them. There was a direct relationship between how much time the dysphasic children had to process brief, rapidly successive information into their brains and how well they could respond.

I replicated this finding twice and got the results published in *Nature*—my first scientific paper as a first author. At the time I didn't even know that *Nature* was a special journal; I had just noticed that *Nature* papers were much shorter, so I thought I could probably write one of those more easily. I sent off the paper and it got accepted without any revisions. It was only later that I realized that it is a big deal to get a paper published in *Nature* or in *Science*. At any rate, that started me on my career path.

Since then your research has expanded, both in the fundamental direction and in the applied direction. What are some of the burning questions in your current work?

Actually, my work has gotten quite controversial in the last several years, so that will get us into some of the other questions that I'm sure will interest you. I have been doing research on this topic for over thirty years now. What we have found through many further studies is that children with delayed language also have trouble processing nonlinguistic, sensory information. Just about anything you give these kids, even in other sensory modalities, isn't processed very well if it is too brief or in too rapid a succession. We did different studies with different kinds of signals—changing the duration of the signals, the intervals between the signals, and the number of the signals in a row. The time window turns out to be really critical for speech perception. Normally developing young children need only about ten milliseconds between two short tones of different frequency to distinguish differences in their tone and order. Language learning–impaired children—which is now an umbrella term for kids with developmental language problems, who will also usually grow up to have reading problems and other kinds of learning disabilities—need about three hundred milliseconds. This is a huge difference in brain processing time, a factor of thirty. It just knocks your socks off when you think of the potential developmental consequences. It helps to explain many of the attention and memory problems that characterize these children.

So some of the attention and memory problems are really processing problems?

Yes. You can't remember something you can't encode or process. Everybody thought this finding was very interesting. But so what? These kids have speech and language problems. If you can't process acoustic information well, you are going to have difficulty learning speech.

You might think that's a very straightforward, obvious extension. But the most prominent theories in the field, then and now, have built on the linguistic theory of Noam Chomsky, which argues that humans have an innate, language-specific module in the brain. The hard line of this theory is that "speech is special," that it does not have to be learned and is not built up by stringing together bits of acoustic information. Over many years research has indicated that speech is processed differently from other auditory signals and exclusively so by humans. So the models of basic auditory processing that I have studied, especially those that were derived primarily from animal research, could have nothing to do with the development of speech perception in humans.

Now here is the more critical point, and this is the part of my theory that is more controversial. Our hypothesis is that if you can't process rapidly successive sensory stimuli well, then you will find it very difficult to process the rapid acoustic changes that characterize speech. The problem for language learning–impaired children, we believe, is the sensory rate of integration, which is much more basic and pervasive than speech alone. The problem lies in the rate at which the brain can integrate brief, rapidly successive input in the auditory modality.

What I did, theoretically and experimentally, was to link nonlinguistic acoustic processing problems directly to developmental speech and language deficits. In order to do that, one has to look at a speech signal as an acoustic event and ask, "How does the brain process the acoustic signal and make it into speech?" I hypothesized that what differentiates speech from other acoustic signals is primarily the rate at which frequency changes occur within the ongoing stream of speech. These are very, very rapid—among the most rapid processing the human brain has to do. The brain somehow has to segment this ongoing stream of sound, find consistencies in it, and learn to represent those patterns as the speech sounds, or phonemes, of our language—all in order to build words, phrases, and sentences from them.

Does this mean that the rate of acoustic change is the key signal for the brain to recognize that a stream of sound should be processed as speech?

It's at least one of the key signals. The rate of change will certainly be a limiting factor in learning speech if your brain processes more slowly. There are also frequency cues, intensity or amplitude cues, and rapid frequency and amplitude modulations; we know that we have neural maps for all these. We then asked: "Would it matter to speech and language development if you had a roadblock in one of these areas?" We know it matters if you have a hearing impairment—that's the intensity or amplitude problem. If you need more amplitude than you get on a regular basis, then you are going to have a hard time learning to talk.

What my work suggests is that the same may be true of temporal or spectral aspects of sound. If your brain integrates sensory information at a slower rate, then you're going to have to "chunk together" larger segments of information, missing out on the fine-grained details of speech at the phoneme level. Keep in mind that language-impaired children need about three hundred milliseconds, while normally developing children need about ten milliseconds to process two brief, successive frequencies. Well, it turns out that many of the differences between the speech sounds of human languages, like *ba* and *da,* occur in only about forty milliseconds.

My hypothesis was that children who, for whatever reason, have difficulty in processing rapidly successive acoustic cues are also going to have difficulty setting up sharp, categorical neural representations for the phonemes of their language. Think about the infant lying in her crib. Mommy is saying, "What a beautiful baby you are. Look at you!" What does the baby's brain have to do? It's got to process that ever-changing acoustic signal and try to pull out sound patterns that happen over and over again. That's how it sets up its neural map for the coding of speech signals.

Since we don't know which language we are born into, this neural mapping has to be a learned process. What's so exciting about languages, and about this hypothesis, is that it is potentially universal. Different languages have different phonemes, but those phonemes must build the language. If you have a processing constraint that changes the way in which your brain sets up phonemic information—as a dysphasic child does—you are going to approach the whole language-learning progression very differently.

What's fascinating about your hypothesis is that it begins with a pathology, but it leads toward a fundamental theory of speech acquisition.

Yes, it does, and that's the reason that it has become so controversial.

Can you spell out the controversy?

Our theory proposes that basic nonlinguistic processing capacities affect language development. This is fundamentally different from Chomsky's theory that language is innate. In Chomsky's theory you don't have to learn language, and you don't have to be taught it; you come into the world with innate language universals, and you just have to tune them up, or turn them on, for language to unfold automatically. Chomsky's view has driven the field of linguistics as well as the field of speech science. There have been decades of studies aimed at proving that "speech is special," at showing that speech perception is fundamentally different from nonlinguistic auditory processing.

My research supports a different view, that speech processing derives from and utilizes mechanisms common for complex acoustic analysis. For example, we have looked at hemispheric specialization—the differences between the left and right sides of the brain that led many people to believe that an innate "language module" resides in the left brain. Our work suggests that hemispheric specialization may not be linguistic per se, at least not at the phoneme level. Instead, it may be that the left hemisphere just processes and integrates information more quickly than the right and that it is this specialization that makes it ideal for processing the rapidly changing, complex acoustic signals that characterize speech.

Yours is a bottom-up model of speech acquisition, which is in striking contrast to a top-down model of preexisting language structures like Chomsky's. Does your theory have any implications for how we humans view ourselves in relationship to other higher primates?

Yes, that is at the heart of the issue, and I believe that's why it has become so emotional and controversial.

If humans alone had Chomsky's language module, we would be Aristotle's "rational animal," unique among the animals. But if our human language ability is mostly learned pattern recognition, then we might just be a little quicker and more discerning at it than our primate cousins. The line between us and the higher primates would be a matter of quantitative degree rather than qualitative difference. Is that where your model is pointing?

Basically, yes, but I do not think that this way of viewing speech acquisition precludes qualitative differences as well—including, say, having a soul. I touched on these issues in a paper I was invited to write for the Library of Congress, a commentary on changes in the field of psychology in the twentieth century and advances that might take place in the twenty-first century. Although I believe it was necessary to create dichotomies—such as "nature versus nurture" or "mind versus soul"—in order to develop psychology as a science, we must now move beyond these dichotomies if we are to progress. The likelihood is that specialized modules for language have developed over time in humans. But the data, as I read them, suggest that the modules are indeed dependent on more basic neural processes that are common to lower species as well. The data also strongly suggest that learning from the environment must interact with innate neural substrates, and this interaction is what creates each individual brain.

In your view, are the specialized language modules subsequent to the basic neural processing capacities?

Yes, we can look at this in terms of evolution. We've done about twenty-five years of research on the etiology, the origins of language and reading problems in humans, and also on hemispheric specializations. We recently developed an animal model to investigate whether animals might, like humans, show hemispheric specialization—not for speech sounds but for rapid acoustic changes of sounds that occur at similar rates to speech. We designed a dichotic listening task for rats. Now there have been hundreds of studies showing that humans have what's called a "right ear advantage for speech." Our work has shown that this right ear advantage—although it serves speech—is not selectively for speech. Rather, it occurs for any stimuli, nonlinguistic as well as linguistic, that change at the rates at which speech changes. Furthermore, it is not exclusive to humans, as has been claimed: rats also show a right ear advantage for rapid acoustic processing.

Most recently, we have been working on an animal model for dyslexia, based on some anatomical differences that Al Galaburda and his colleagues found in the brains of dyslexics. We started collaborating with them. It turns out that you can induce similar structural changes in the brains of rat pups. We found that these rats, with the same anatomical and physiological differences seen in the brains of dyslexics, also have the same rate-processing deficits for auditory information.

Can you tell us now about your current work on remediation?

Sure. We hypothesized that, if our theories are correct, some forms of language-learning problems result from difficulties in rapidly integrating acoustic information. Thus, if we could increase the rate at which these children process acoustic information, this should improve their language development. As a result our most current work has begun to focus on remediation.

In 1993 I started collaborating with Mike Merzenich, and this wonderful collaboration has dramatically changed both of our lives, both scientifically and personally. Mike's work on neural plasticity had shown that you could use experiential training to alter sensory neural maps in the brain, as well as to speed up the rate of neural processing. At a meeting at the Santa Fe Institute we both had this major "ah-hah" moment. We thought, well, if you can alter the sensory map at the cellular level, could you speed up the rate of processing in children with language disorders? If you could speed up the basic rate of acoustic processing, then theoretically you could improve these children's ability to extract the rapid acoustic changes within speech. And if our hypothesis was correct, that should lead to improved language abilities.

Meanwhile, for many years I had been interested in using computer technology to slow down the rates of acoustic transitions within the speech wave form. In fact, way back when I was doing my Ph.D. thesis, I had used computers to expand the duration over which those critical forty-millisecond frequency changes occurred in speech. I had used a computer to stretch out the acoustic transitions that marked the difference between sounds like *ba* and *da* from forty to eighty milliseconds and then to make the vowel shorter, keeping the whole thing in real time. The language-impaired kids showed tremendous improvement in their ability to discriminate between these acoustically modified speech sounds with extended-duration transitions.

That's amazing. In some cases the therapy could take place in real time, rather than at a vastly slowed-down rate.

That was the idea. But making these precise acoustic changes at the single-phoneme level was hard enough. Making them in real time, in ongoing speech, was something beyond my technical capability. When Mike and I got together, his primary interest was to do a neural plasticity study with language-impaired children to see if we could speed up their neural processing rate. My primary interest at the time was to develop a "speech prosthesis": essentially, a central auditory hearing aid that slowed down the acoustics within speech sounds in real time for these kids. We decided we would try to unite these two approaches, because both should be important. We did a series of studies for the first time in the summer of 1994, and the results were spectacular. The results showed such enormous improvement in language comprehension for these kids that we couldn't believe it. Mike just couldn't wait to publish it in *Science*. But I thought, "Whoa, wait a minute, we have to replicate this with well-matched controls first"—because I knew this would be extremely controversial.

Why would it be controversial?

A lot of people who are not in the field don't understand that to this day. However, within the field of linguistics, language is sacrosanct. The prevailing belief is that language is innate to humans and is encapsulated in a special language module. It is innately acquired, not explicitly taught or learned. We, on the other hand, are saying that we can alter language acquisition through training. So we are really stepping on a lot of theoretical toes.

Another part of Mike's work, driven home by our collaborative work, went against the current wisdom about critical periods of development. The children in our initial studies were five to fourteen years old—way past the critical period for language development, which is supposed to take place in the first several years. In subsequent field studies we have found similar results into adulthood. Yet we were getting improvements of about a year to three years in language abilities in just six weeks of training. How in the world could that happen?

We've been led to believe that if the critical period of language development is passed, you cannot change it any longer. Are you saying that you have solid empirical evidence that those critical periods are not so critical after all?

Yes, that is what the results of our training studies imply. We are regularly able to get improvements of one to three years in language ability in four to eight weeks of our training, at an age where it should be impossible to gain any significant improvement in such a brief time. Usually, the longitudinal studies have indicated that it takes approximately two years of speech therapy for every one year of language improvement. But with our Fast ForWord training programs, the improvements come at a much faster rate.

Well, you may be stepping on theoretical toes, but nobody could complain about dramatic improvements in these children's language skills.

Mike and I thought so too. We thought everyone would be thrilled, but of course those scientists who do not agree with our theories discount our results.

Because the standard theories say that improvements don't occur at this pace?

When results go against long-held beliefs and expectations, they are very difficult for some to accept right away. This is natural.

We've now developed computer games to help speed up the rate of processing with nonlinguistic signals, but they also use acoustically modified speech. To do that we've developed an algorithm that alters those parts of ongoing speech that change rapidly (anywhere between three and thirty Hertz), making those parts longer and louder. Right off the bat we found that it worked in a dramatic way. We then did a second, controlled study in the laboratory with language-impaired kids and again saw a year and a half to two years' improvement in language skills in about four weeks.

We published two papers in the same issue of *Science* and ended up with an enormous amount of press publicity. As you can imagine, that resulted in about thirty thousand phone calls into Rutgers University and the University of California–San Francisco—it actually shut down the phone system. It was overwhelming. Most of the responses have been enormously positive. Parents and teachers and many colleagues were excited and wanted us to "send us the CD." We didn't have marketable software at that point, just a series of laboratory studies.

It was these results that threw you into your work on therapeutic applications?

Yes. After the first summer, when we realized that we had something pretty amazing, we had filed patent disclosures with the university. As the power of this new approach became clear, the real issue became how best to get this to the children who could benefit. Of course, as idealists we wanted to get it out, immediately, to all the millions of kids who had language and reading problems. Mike's mantra became "faster, faster."

Mike and I cofounded Scientific Learning Corporation with two other colleagues and raised venture capital funding to start it up. It is now a thriving business with over three hundred employees, and is traded on the NASDAQ exchange. To date, over fifty-five thousand children have completed Fast ForWord training, and our programs are currently in use at over two thousand public schools in the United States.

How has that changed your life as a scientist?

In every way.

Do you consider your commercial success a benefit or a cost?

Both. I own a lot of shares in this company, and when the company went public, I became publicly identified as a founder. This has significantly compromised my ability to be seen as an objective scientist. Our financial interests in the company became a clear conflict of interest for both of us—probably more so for me than for Mike, since my scientific work had always been in the domain of language impairment and in particular this acoustic processing hypothesis.

Let me ask, then, about your advocacy role. As objective as you try to be as a scientist, you are also an advocate for a therapy that has the potential to improve the quality of life for many children. That reminds me of Jane Goodall's work: beginning as a scientist, she then found that she needed to defend the chimps and their habitat. When she became a spokesperson for the animals, she was sometimes attacked for trying to continue her scientific work.

Yes, it's very much the same for me. I have become a spokesperson for speech, language, and reading problems in children: I've had a lot of press interviews, and I generally get called for comment when anything comes up in that area. At the same time we have a commercially available therapy based on some theories that not everyone agrees upon.

Some people would say that science is defined by objectivity and detachment from the results. Others would say that when our science offers the ability to improve the quality of human life, there is an overarching obligation to do so—even at a certain cost to objectivity. I think I know which side you would come down on. Is there a resulting loss in your ability to do research science?

Rather than thinking about it in terms of losses or gains, I prefer to think about it as a transformation in my vision of what I can ultimately achieve with the gifts I have been given. I have always felt, at a spiritual level, that one's purpose for being here is to discover the gifts you were given and then figure out how to give them back to others. I always thought I was doing this best by being a good scientist. But if you're fortunate enough to have your science become of practical use, then the obligation is greater to try to see that the practical use is handled appropriately, that it is as efficacious and beneficial as it can be to as many as can benefit.

Is there something wrong with the view that science is supposed to be "pure" and detached?

I wouldn't say there is something wrong with that view, but I think it may be more of an ideal than a reality these days. For better or for worse, science in general has become big business. Very few of us can do our work without getting very large amounts of money to support our labs. Because of this, pure science is conflicted at a lot of different levels. In society we see business as being very separate from science, but as the scientific enterprise has grown, it has become less like the ivory tower and more like business.

In the medical sciences the goal of enhancing life or reducing suffering is always at the center. A given individual can move from more theoretical work to clinical or therapeutic work and even to advocacy. Is that unusual within science as a whole?

It is different in the medical sciences, I would say. That certainly proved to be true for me. I don't think that I have written any grants where I didn't say, and truly believe, that the goal was to improve assessment and treatment for children with language disorders. However, although I am also trained as a psychotherapist, my scientific work was always separate from my clinical work. I had never done remediation research or treatment research; I'd always done basic theoretical and etiological research. I've primarily seen the work with language-impaired children as a means to the end of understanding more about how the brain processes speech and language, with hopes that this would lead to improved assessments for language disorders.

But then your theoretical advances on brain processing turned out to be a means to the end of a life-enhancing therapy.

Right. That took me by surprise, actually—I think that Mike probably saw the therapeutic potential more than I did. First of all, I frankly didn't think that they were going to work; I was a skeptic. Like others, I thought that these children had passed the critical period and that we were not going to be able to speed them up.

I never thought we were going to be able to attack the problem head-on; I thought we were going to have to work around it. I'm not sure we have found a way to "fix" the problem, but we have certainly speeded these kids up. We changed "slow" kids into "quick" kids. Not as quick as other kids, but we made about a threefold improvement, and all in six or so weeks—and, most important, their language went up commensurably. I never thought this would be possible, so I certainly understand our critics' disbelief.

Of course, it's possible that our theories and explanation about what is really happening in the brain will someday be shown to be wrong. But when you see the immense practical benefits, you're almost tempted to say, who cares? The kids are getting better! This is where the real conflict of interest arises, because the scientist in me is supposed to care most of all about *why* this is happening. Yet I care first and foremost *that* the kids are getting better. Only next do I care about understanding exactly what is driving these changes—and now less for the sake of the theory than for the sake of becoming more effective in improving these kids' lives. My focus has changed: unraveling the workings of the auditory system in the brain still matters, but whether my theory is right or wrong, or who gets the scientific credit, is less important to me. The most pressing issues for me now are to develop additional applications and to assure that these advances get to as many people as possible.

It's almost as though there is a calling in your research, a calling to be involved in the therapeutic application of what you've learned about brain processing.

Yes, absolutely. If these new approaches are this powerful for children with language impairments, we need not stop there. The research tells us something very basic about how we can alter the way the brain processes information, and it certainly need not stop with language impairment. In the language domain you can imagine using similar techniques for teaching English as a second language or for reducing foreign language accents or—coming back full circle in my work—for remediating adult aphasia. In the motor domain there are efforts now, using adaptive motor training, to improve walking and dexterity in patients with stroke or spinal cord injury. Even in the emotional or psychiatric domains, you can think about new approaches that modify the brain's processing of information at a much more basic physiological level. Brain plasticity training has enormous potential as a whole new form of therapy.

It seems to me that your results might lead to some fundamental changes in how we conceive the human person and the relationship of humans to animals. Would you be willing to speculate about these implications? It is, I admit, more of a philosophical question than a scientific one.

It can also be a scientific question; there is a whole field of research on the evolution of the nervous system. There is no question that humans can do different things than can any other animals, but there is also no question that some higher primates can do some things better than some lower primates, and some lower primates can do some things better than animals below them. So the question is not whether there have been evolutionary changes to the nervous system. Rather, the question is whether these constitute a systematic enhancement of certain kinds of functions for adaptation in the world and whether or not language and other higher cognitive functions grew out of those basic systems. Or whether, alternatively, language really did spring full blown, as it were, from the head of Zeus.

But the latter is the position that your work challenges.

Yes, my work does challenge that position because I suggest that there are some basic mechanisms that must be functional at certain levels before you can unfold language normally. But the same is true of hearing and cognition. We used to call people "deaf and dumb," before we finally recognized that deaf people couldn't learn oral language, which was the reason that they appeared to others to be retarded. Remarkably, there remained great resistance to teaching language via manual signs. Although we want to improve human outcomes, we as scientists are very cautious by nature, which also may lead us to resist rapid change.

What about the perennial question of the soul—an ontological difference that distinguishes humankind from other animals?

I haven't thought much about whether brain research could address the issue of the soul. But let's say we could get animals to process as quickly as children, and then let's say we were able to immerse them in language from the earliest stages. Would those animals eventually know how to talk? First of all, they won't be able to talk orally, because they don't have the physiological mechanisms for articulation. However, studies have been done with sign language in apes, and my sense is that they have been remarkably successful; given the challenges, that apes can learn as much as they can is quite remarkable. But, of course, linguists say that's not language at all.

People who have been fond of a pet would not question whether they are communicating with the pet at an emotional level. They talk to that pet and love it. I don't say that the pet understands their language at the level of linguistics—but certainly there are connections between humans and animals that are quite profound. The emotional part of that relationship can be transforming; it can be human. As a psychotherapist I often "prescribe" that my severely depressed patients get a dog or a cat as a treatment, because it can be a stepping-stone to getting back into contact with humans.

Do you see any connections between your path as a scientist and an advocate on the one hand and your interest in spirituality on the other?

I've always felt that there was a connection between my spiritual life and my career. An axiom that has always been important for me to remember is, What I do and who I am are not the same thing. This is a struggle for most scientists, whose career defines their persona to other people, and it has certainly been difficult for me. The spiritual quest for me has been a struggle to keep a "me-ness" separate from my science. As my advocacy role has grown and I have become more publicly visible, it's become difficult to keep that sense of who I am as distinct from what I do.

Clearly, I've always been interested in therapy, because I also took time out from my science to get licensed as a clinical psychotherapist and to see patients. I've always felt that having the human contact was really important, and getting more involved in therapy has brought me a lot closer to people.

I also chose to do research that brought me in contact with children and their parents. As a scientist I always felt a separation; when I gave scientific talks on language impairment in children, I wasn't talking directly to the parents or the kids. But since we started providing the Fast ForWord therapy, I've gotten many more requests to talk directly to the people that our therapy helps. Frankly, it's a challenge to balance it all.

Is your willingness to do some of those talks, your desire to help, motivated by spiritual concerns?

Honestly, at the beginning I was motivated by the necessity to get information out about our new remediation. We needed to do a large field study that demonstrated that the laboratory results could be replicated in clinics and schools, so one of my first jobs for Scientific Learning Corporation was to train therapists. Getting results in the lab and getting them in the classroom are two different things; in order to assess the clinical efficacy of Fast ForWord outside the lab, we had to teach therapists to use the program and be sure they were using it correctly. In a sense, my job was to go out and train hundreds of therapists as research assistants. I did thirty-three training seminars all over the country in the first year, and I got pretty burned out doing that. But, subsequently, I have truly enjoyed the personal contacts with parents and therapists. I have had the pleasure of having so many tell me that our work has changed their child's life. Many of them, mothers and fathers, have broken down in tears when talking to me. I can't describe the feeling—there is nothing like it.

This is one of the reasons why I ended up providing some individual therapy as well. Freud said that the definition of maturity was giving up immediate gratification for long-term goals. Well, scientists have to be the most mature people in the world, because it's years and years for us to get any gratification out of our science—if we're lucky. But as a therapist you get that sense that you are helping immediately, or at least you get feedback if you're not. The same is true with developing these new remediation programs. When I walk into a school and see kids in so many classrooms doing Fast ForWord training programs, the feeling goes well beyond anything I ever imagined I would be able to accomplish in life.

Yes, that is a powerful experience. I note that, while you've spoken of other tensions in your work and life, I've not heard you describe any conflict between your science and your spirituality. Did I miss something?

No, I've never seen it that way. I don't have a formal religion. If I've had a conflict, it is in finding my spirituality through a formal religious service—although I love the music. I was brought up Jewish and I converted to Christianity, because it seemed an obvious extension to me. I've never seen a conflict between those, either. I think that the spiritual world is an integral part of the physical world, and as long as one doesn't try to use religion like a hammer to impose rules or to stifle personal expression or creativity, then there is no division. I certainly have never seen anything in the Bible or any other direct religious teaching that is in conflict with that. I do have a daily dialogue with God. This has been part of my life from a very young age, but it is more of a conversation than something that I do formally in a special place. My main goal in life, both as a spiritual person and as a scientist, is the same: to use whatever gifts, whatever love God has expressed in me, to the best of my ability—and to share these gifts with others as much as I can.

11. Henry Thompson

Henry S. Thompson is reader in artificial intelligence and cognitive science in the Division of Informatics at the University of Edinburgh, where he is based in the Language Technology Group of the Human Communication Research Centre. He also is managing director of Markup Technology Ltd., an XML-related startup. He received his Ph.D. in linguistics from the University of California–Berkeley in 1980. His university education was divided between linguistics and computer science, in which he also holds a master's. While still at Berkeley he was affiliated with the Natural Language Research Group at the Xerox Palo Alto Research Center. His research interests are in the area of natural language and speech processing, which he

---

Gordy Slack interviewed Henry Thompson; Zach Simpson edited the interview for publication.

approaches from the perspectives of both applications and cognitive science. His work thus ranges across the fields of natural language parsing, speech recognition, machine translation evaluation, the fine structure of human-human dialogue, language resource creation, and architectures for linguistic annotation.

Since coming to Edinburgh in 1980, Thompson has become a leading member of the British and European speech- and language-processing research community. His current research is focused on articulating and extending the architectures of XML. He was a member of the SGML Working Group of the World Wide Web Consortium, which designed XML; is the author of the XED, the first free XML instance editor; and is coauthor of the LT XML toolkit. He is a member of the XSL and XML Schema Working Groups of the W3C, holds a World Wide Web Consortium Fellowship, and is lead editor of the Structures part of the XML Schema W3C Recommendation, for which he cowrote the first publicly available implementation, XSV. Thompson has presented numerous papers and tutorials on SGML, DSSSL, XML, XSL, and XML schemas in both industrial and public settings since the mid-1990s.

He attended Quaker schools in the Philadelphia area and is a long-time member of the Religious Society of Friends. He is keenly interested in ethical issues arising out of the sciences in general and computer science in particular. One hears a clear spiritual motivation in his concern for "right action," though he devotes few words to doctrinal disputes and details. In this emphasis, and perhaps also in the pregnant pauses for reflection, one senses the hours of silent worship that are a major spiritual practice within his tradition.

Henry Thompson is a high-energy discussion partner. Just as his professional work moves among computer science, computational theory, and linguistics, his conversation moves smoothly back and forth between facts and values, in a manner that one rarely finds within the sciences. Thompson seems equally comfortable discussing his science, the ethical issues it raises, and the religious context to which he turns for answers. This conversation took place in 1997.

Welcome. To begin with, Henry, would you please say a few words about your religious background?

Religiously speaking, I grew up in a suburban Philadelphia home in which my parents were, roughly speaking, nonpracticing atheists. They were a part of the generation that fought the battle not to be terribly concerned about religious observance. Although I was baptized, we never went to church on Sundays. I have clear memories of going to an ordinary church service for the first time when I was thirteen or fourteen, when I was visiting some friends. But when I was in fourth grade, I changed from the local suburban public school that I was attending to a Quaker school, Germantown Friends School in Germantown, Philadelphia. By the time I left there at seventeen and three-quarters, I considered myself a Quaker—not only because it's part of the life of most Quaker schools but because, by the age of sixteen, I had made a conscious decision to attend worship meetings on Sundays. But when I went out to Berkeley for university in 1968, I attended only sporadically. I didn't resume attendance at the weekly meetings until I moved to Edinburgh sixteen years ago. Not too long after that, I formally applied for membership and was accepted as a member of London Yearly Meeting of the Religious Society of Friends, or Quakers. So I've loosely been a Quaker off and on all my adult life but only formally for the last twelve or fourteen years. I am an active member of my meeting; virtually every Sunday I go to meeting for worship, I served as an elder for six years, and I play in the meeting *ceilidh* band. [*Ceilidh,* pronounced "kay-lee," is typically a party or gathering in Scottish communities.]

Have there been any other movements that have informed your spiritual trajectory?

Interestingly, because my wife is Dominican Catholic, Catholic theology has been a significant influence on my theology and my religious perspective. So our children on most Sundays will go to mass and to meeting, and therefore I continue to be actively influenced by Dominican Catholicism. In living between these two traditions, I find that there is actually a surprising amount in common between Catholicism and Quakerism. They sort of meet around the back, leaving the other Protestant denominations in a different category. And I think they both recognize this roundabout way of theological agreement. It's not just the Quakers that think this; the Catholics think so too.

So that's the basis for my religious perspective. It also happens to be the reason why I don't use my title, Doctor, because, for complicated historical reasons stemming from the seventeenth century, Quakers have not used titles since then. It has to do with having been sent to jail for refusing to take their hats off in front of the king at that time. I try to remind myself about that testimony by not referring to myself or to other people with titles. I'm not certain if that's what you meant by "your spiritual trajectory." If so, it's this blend of two mutually appreciating traditions that composes my spiritual trajectory and shapes my life.

I'm interested in your turn to Quakerism as a teenager. Was it just a gradual deepening of interest? Or was there what you would describe as a spiritual event of any kind?

It certainly was gradual. It's probably not surprising that conversion experiences are not high on the list of things that Quakers talk about, because of the imperative upon silence in worship. Silence is at the heart of the Quaker approach to spiritual discipline in general and worship in particular. Certainly, for me, what happened was a growing realization of the importance of silent meeting. I was struck by these little old ladies who came in from outside school on a Thursday morning when we had meeting for worship; [they] weren't teachers necessarily but were members of the meeting that had oversight of the school. Every Thursday they would stand up and tell the truth in ways that I found compelling—ways that I couldn't address in any way other than to take them, and their beliefs, quite seriously. Their teaching and testimony had tremendous gravity, and it compelled me to really deliberate over the weight and power of their message and their beliefs.

So it was the more intellectual side that drew you to Quakerism?

There is certainly room for a more intellectual approach within Quakerism that has definite appeal, especially given my analytical nature. I found the direction of Quaker theology to be much more congenial to my more scientific spirit. I ended up in linguistics and cognitive science and artificial intelligence, all of which are sciences, as various people have said, that "pay you to take out your brains and play with them." And so an inarticulate conversion experience or ecstatic Pentecostalism was never really attractive to me. I think that's also why I've found the Dominicans so congenial, because, along with the Jesuits, they represent the intellectual wing of the Catholic Church.

Not that this intellectualism is devoid of a spiritualism. As I have matured, experiences in meeting for worship have given me a sense of the Divine Presence in a way that I wouldn't describe as a thunderclap but that has validated my belief or feeling that there is more to things than we see in the physical world. This experience is expressed in the Quaker phrase "the gathered meeting." After a meeting for worship that has really worked, people stand up and look at each other and say, "Oh, there was something going on there, wasn't there?" Experiences like this involve, at least in a small way, a breaking through into the real world of the transcendental, of the ineffable. Such moments show the deeper and more poignant elements that underlie all of our reality. It's this maturing recognition, both intellectual and spiritual, that has drawn me into the more silent and contemplative traditions.

How does this emphasis on silence and the intellectual dimension affect your practice?

In Quakerism there is certainly a pervasive emphasis on walking the walk as well as talking the talk. There is a particular Quaker injunction that says, "Let your lives speak." For a Quaker, if you want to understand the Christian message, you focus on how to live your life. Watching what people do at a Quaker meeting is a very useful, practical reminder. Another way of putting this point is to say that, like Judaism, and unlike almost any other Western religion, Quakerism is an *orthoprax* sect rather than an orthodox sect. What matters is what you do. There are virtually no beliefs held in common among all late twentieth-century Quakers, but they all go to meeting for worship on Sunday and sit quietly and say what they have to say if they feel bidden to do so. That's what they share.

Most late twentieth-century Quakers do not hold to a common core of beliefs, but they do at least affirm the Quaker way of living and being. What we share, then, is this common emphasis on living and the maintenance of community through the meeting.

I was talking to Arno Penzias [the now-retired chief scientist at Bell Laboratories] last week, who said, "When you ask a Jew if he wants to become a Christian, he says, 'What do I need to do?' And when you ask a Christian if he wants to become a Jew, he says, 'What do I need to believe?'"

I think that characterizes the Quaker way of life just as well as the Jewish. The emphasis is always on what one does, how one lives. Actually, I got the word *orthopraxy* from a close friend of mine, Mitch Marcus, who is a serious practicing Jew, a thoughtful one with whom I have had many conversations. For him the question remains the same: How do I live? I believe it was Kant who observed that, in life, there are two basic questions: Why is there something, rather than nothing? And, How ought I to live my life? Quakers typically have little interest in, and little to say about, the first question, but they have a great deal to say about the second. Aside from their disinterest in existing established religions and churches, the thing that most distinguished the founding Quakers of the seventeenth century was that they believed in the perfectibility of humankind and the possibility of the kingdom of God on Earth. And they were therefore mixed up at that time with people like the Ranters and the Levellers and the Diggers, who were persecuted relatively indiscriminately along with the rest of them. From the perspective of the Quakers this was grossly unfair, as there were some fairly important differences between them. One of the things one says about going to meeting for worship on Sunday is that one is practicing living the kingdom. Likewise, we refer in the same way, if somewhat grandiosely, to the means by which we manage our church. One of the main aspects of Quakerism is that there is no laity. Outsiders say there's no clergy, but we like to think about it the other way around. What that means is that we run the place ourselves; we call it our "gospel order." There is a drive to moral and spiritual betterment that coincides with being a Quaker. It affects the way in which we try to run our meetings qua churches, and the institution goes by that name, although it's being a bit generous in various ways. But all this doesn't come at the expense of humility. The standing joke is that somebody who has heard about Quakers is asked, "Would you be interested in becoming a Quaker?" And they say, "Oh, I couldn't possibly; I'm not good enough." And the Quaker response is, "Neither are the rest of us. We're just practicing."

In what ways does your profession in information science affect your practice as a Quaker, your orthopraxy? Are there points of conflict?

There are moments in the work I do that compel me to make ethical decisions, and these ethical decisions draw into relief the tension between practical needs and a living religious sense. I'll take the liberty of giving you an example. In the late 1970s and early '80s I was involved in a movement called Computer Professionals for Social Responsibility, or CPSR; I helped start a similar group in the United Kingdom when I moved here in 1980. CPSR arose largely out of concern about the dominance of military funding for computer science research in the United States. Especially in the European context it subsequently came to focus on issues having to do with "launch on warning." Launch on warning was a particularly poignant topic here because of the political dynamic. The Americans were saying, "We have missiles in the eastern parts of Western Europe. The Soviets have missiles in the western parts of Eastern Europe. The missile flight times are such that their locations are only fifteen minutes apart. What that means is there is no possible human mechanism that can be given the responsibility for getting those missiles off the launch pads in time to avoid destruction in case of a Soviet attack. It therefore follows that we must put our missiles under automatic controls to be a credible deterrent." In short, the political dynamic mandated what the technical necessity was—despite the fact that it obviously would have been irresponsible for a computer scientist to put what were essentially fallible computer systems in control of nuclear weapons. One of our goals as an organization was to show the recklessness of this action, that is, to show the inherent ethical flaws in basing a strategic response on an automatic computer system.

And was this appeal effective?

We were actually fairly effective in getting computer scientists in this country to say that they did not want any part of the proffered U.S. funding for SDI [the Strategic Defense Initiative, also known as Star

Wars] research in this country. Here we were reasonably successful. We managed to get a pretty unanimous declaration of disinclination on the part of the British technical community to have anything to do with "launch on warning" computer systems. In a more general sense what that all led up to was a concern with the moral dimension of empowering computational systems to act in the human realm. Although this is an issue that remains almost completely ignored, it seems to me to be an issue of profound moral and practical significance. I am particularly interested in instances, such as the one raised by the SDI plans, in which there is a major rational empowerment of an artificially intelligent computational system in a significant human context. Another excellent example concerns the judicial system. I have heard people say, "In the future we will have 'computational judges' who aren't tired and don't get bored and who will render fair and just judicial verdicts with utter reliability." Leaving aside the immediate worry that artificial intelligence just isn't up to such responsibility, and won't be in the foreseeable future, there are obvious moral questions about such an example that have been almost completely neglected. I think even the possibility of such a situation raises considerable moral dilemmas that I'm not sure we're prepared to address or even face.

What kinds of issues do you think are raised by this possibility—the possibility of artificial intelligence's operating in the realm of legal judgments, of justice?

Well, in the example of a judge, there needs to be consideration of what is involved in being a judge at all—being one who renders judgment. This includes, from the perspective of the accused or someone who is powerless, an analysis of what it means to submit oneself to judgment. What does it mean, more generally, to place yourself into others' hands, in circumstances where they have been given responsibility for a significant aspect of your life—whether it's literally your life, as in the case of doctors, or your education or your children's education or how you are handled in the legal process? I think it's essential to have this recognition in a judge and the empathy that goes along with it. My submitting to judgment depends, in no small part, on my recognition that the judge takes responsibility for his actions.

Responsibility—self-conscious responsibility—is fundamentally a moral category. It's not just an "imputed" responsibility; it requires a direct admission of authority and the recognition by the participants of the ways in which they are responsible for what occurs. A teacher's consciousness of his responsibility for the children that he teaches, or a judge's conscious recognition of her responsibility for the judgments that she renders and for the people upon whom she renders them, are absolutely constitutive of those roles. If that's true, then reflecting on the origins of this self-conscious responsibility is eventually going to be necessary—especially if we start to design systems capable of some form of moral problem solving. And at that point, if not before, a thoroughgoing understanding of the grounding of our moral behavior in our religious sensibilities becomes necessary. I ultimately cannot divorce the issues of conscious moral responsibility, as in the case of a judge or a teacher, from their religious implications.

If, even in the religious life, what is really most important is to behave properly, then wouldn't it be sufficient for a computer judge just to behave properly, even if it does not believe it is responsible for those judgments? Is there a necessary link between believing I have a moral responsibility and actually doing the right thing, that is, being a good judge?

That's a very good question. Your question is critical for Quakers because of our emphasis on right action. When George Fox [one of the principal founders of Quakerism] spoke of the possibility of living in the kingdom, he based his belief in part on the saying attributed to Jesus, "Be ye therefore perfect." Let me start with the well-known anecdote. One is asked, "How do you paint perfectly?" to which one responds, "That's easy: just become perfect and paint naturally." Now most of us aren't perfect. What is crucial to trying to live well is the self-conscious process of deliberation, of weighing options, which I think constitutes responsible moral action. This doesn't mean having to deliberate over everything or even having an exhaustive consciousness of all my moral options. We often rely on role models and on precedent. But the self-conscious process of deliberation still remains central. The joke "If you have to think about something before doing it, does it count?" reveals a naive wish that we could live our moral lives completely naturally, without deliberation, like taking out the garbage or giving change back to a person. But I think most instances of right moral response involve an explicit recognition of one's role and responsibility; they don't just spring naturally from a perfected self.

So how do these considerations bear on the question of an artificially created moral agent?

I think views in this field have evolved considerably over the past twenty years. At first, the mainstream view of artificial intelligence [AI] was of computers or robots whose behavior was designed into them. So if you didn't design into your machine a moral sensibility or deliberative capacity, it wouldn't have one. But in the years since then, the debate has become much more evenly balanced between two dominant perspectives. If I can simplify it a bit, the one side holds that we will build out-of-the-box intelligent artifacts that will spring forth fully formed from their creator's brow, as it were. This was by far the dominant orthodoxy in the late seventies and early eighties. The other side holds that, because the first is too hard, we will build the computational equivalent of a child and let it learn to be an adult. This is easier, because babies are much simpler agents than adults. Of course, this route is pretty hard too, because we still don't know very much about how babies get to be adults in terms of their intelligence—and much less about the development of concrete moral agency.

That transformation itself is probably one of the most complex things that a human will ever do in moving from infancy into adulthood.

Absolutely. So there are a number of questions that arise when one says, "We will create computationally sophisticated but behaviorally trivial artifacts and let them learn to be judges or doctors or whatever else it might be." There are already very simple mechanisms that achieve such a transformation—namely, that seven or eight pounds of protoplasm that starts out life as a newborn baby. But the complexity of that system is already well beyond anything that we currently understand in computer science. I believe that raising questions about the evolution of moral sensibility and responsibility in biological organisms will shed a lot of light on how an artificially constructed system could achieve that level of development. Still, I should say that I don't think the question of the development of moral reasoning is just a biological or theoretical one. It's also a theological question, because it touches on whether or not moral sensibilities are evolved, or given, or both. The view that I find most adequate is Alasdair MacIntyre's answer in *After Virtue*. MacIntyre holds that moral sensibility and agency are both accumulated from the creation of a life narrative. I think this most clearly expresses how moral responsibility and virtues develop over time for most of us.

You raise an interesting point about the evolution of moral responsibility and its connections with theology. Do you think moral development is tied in any way to the development of a spiritual sense? I can see awarding a medal of honor to a machine, but I can't see wanting to baptize it.

Interestingly, it appears that there are various different spiritual bases that induce virtuous behavior. Across a fairly broad spectrum of religious traditions, the consensus about what constitutes a virtuous life is not perfect by any means, but it's much better than chance. It thus seems that, for most people, moral sensibilities are in some way informed by religious considerations. Just the other day I saw a posting on a Quaker Internet newsgroup that said, "When my kids were born, I felt like I needed to get back into church because I felt there was no possibility of their growing up with a reasonable moral sense unless we were going to some kind of church regularly." And that's how this gentleman eventually became a Quaker. But he went on to say, "Since then I've realized that there are plenty of people out there who are leading eminently moral lives without any evident religious sensibilities." There certainly are. So a range of religious backgrounds seem to be possible bases for virtuous behavior. I'm enough of a believer in the value of the spiritual life that I'm going to commit the Catholic sin—and to some extent it's a Quaker sin too—of saying they really are believers, even though they don't think they are. Both the Catholics and the early Quakers tended to do this with respect to other religions, saying, "They're worshiping the same God we are; they just don't know it. They won't acknowledge it, but they really are." This might sound like a somewhat unpleasant sort of triumphalism. But in a more responsible form it's a fairly reasonable basis for what in Quaker circles is known as Quaker universalism, which basically attempts to show the universality of Quaker religious claims. While I'm not completely comfortable with the judgment that we're all latent Quakers or Catholics, I do agree with the view that there is a deep relationship between belief and moral responsibility.

So does this mean that spirituality is a necessary condition for a moral life?

I think the most likely source of virtuous behavior will be one that is guided by some overarching principle. Our virtuous reflex is more easily conditioned when it works from a principled basis. Hence I'm skeptical about the ability of people in general, and scientists in particular, to succeed in the long-standing humanist project of deriving moral behavior from selfish first principles. Without the acknowledgment of something that's bigger than I am, I don't think I could do it. At least I've never observed any attempt to make this derivation that is even remotely convincing. I recently read these words in an article in the *New Yorker* about Arthur Koestler: "We cannot be relied on to do what is right if all we have to work with is our pain or history or interest." So when you ask about artificially constructed moral agents, I have reservations. The development of some spiritual sensibility may be necessary for the creation of artificial intelligence with morality—or at least the recognition of something bigger than I am. I don't know if you can separate the question of morality from that of spirituality.

Let's leave the question of spirit aside for a moment. Today there are many metaphors for the human mind that are borrowed from computational science. How powerful a metaphor or a model for human cognition do you think the computer is?

I've seen it observed, and I agree, that in any given era the dominant metaphor for the brain is drawn from the most complex technical artifact available. The ancient Greek model for the brain was plumbing or waterworks. In the late nineteenth century it was the London Underground. I've actually seen a wonderful illustration from a late nineteenth-century or early twentieth-century encyclopedia, in which the mind is depicted as a very complex railway network. Then for a short time the brain was envisioned as the electrical grid. Now it's computers. Since computers are designed for information processing, it's sort of obvious why they offer a good model. However, many theorists are now advocating an equivalency relation between minds and computers. They don't say, "The mind is like a computer"; they say, "The mind *is* a computer." There is a world of significance in the difference between *like* and *is*. I am convinced that there is a qualitative difference between the two.

How do you envision this qualitative difference? Put differently, what is significant about the recent move to understand the mind more literally as a computer?

What's interesting about this alleged equivalence is that minds are meaning-bearing artifacts. They are systems—complex physical and behavioral structures—that are *about* things, that mean things, that represent the world, and whose behavior is influenced by those representations. Hence there is always a two-way causal connection between the world and our representations of it, and a two-way relation between our representations and our actions in the world. This is the point at which we have to question the analogies between artificial constructs and living minds.

One can ask, as many do, "If you encountered a constructed artifact, as opposed to a biological artifact, that had both of these causal connections—that is, where the world influenced its representations in the way that our perceptual interactions with the world do, and where its representations in turn influenced its behavior in the world, in the same way that our beliefs and desires influence our behavior—then would you have something that was in any significant way different from us?" For proponents of the "strong AI" position, the answer has to be no. However, the question of whether this test would be sufficient for self-consciousness and, flowing out of self-consciousness, for a spiritual life, is not usually discussed. One does read discussions about whether self-consciousness is an intrinsic result of sufficient semantic and computational complexity on the part of any artifact or whether it's epiphenomenal or whether it's biological or mechanical or whether self-consciousness is something we need to understand, like motor processes, in order to be able to create artificial constructs capable of mimicking human behavior. I'm not sure if we'll ever achieve resolution on these fundamental questions. What's generally omitted from the discussions, though, is whether or not such artificial constructs can achieve sufficient complexity, if that's what is really required, to evidence authentic spiritual behavior. Until we have a clearer picture of what constitutes self-consciousness, I'm not sure we can move on to the more interesting question of artificially creating a spiritual being.

So you see the notion of self-consciousness as fundamental to our understanding of spirituality in humans?

I guess I don't believe that an organism, whether biological or mechanical, that wasn't self-conscious could have a spiritual life. And I don't think we can discuss self-consciousness or spirituality apart from questions regarding good and evil or free will. Making choices, especially spiritual choices, necessarily involves self-awareness, which implies free will. Moreover, we normally impute good or evil only to those things that we believe to be self-aware. Notice that one typically doesn't say, "That tree will burn in hell" for having fallen down and crushed a woman in her car yesterday. Nor would we say, "That bacterium will burn in hell for having infected my daughter and killed her." In this way the ideas of good, evil, free will, and choice are all implicated with each other and all feed off each other. In the end there's a sort of circularity here, with all these concepts chasing each other's tails. Thus it doesn't seem obvious to me that merely creating semantically robust artifacts is going to be sufficient to create either self-conscious beings or moral beings. I think semantic robustness is a necessary condition, but it's not enough to yield a meaning-bearing organism capable of self-awareness, free decision-making capacities, and the ability to acknowledge all this.

I wonder too if there isn't a way in which spirituality could be said not just to be reliant on self-consciousness but also on some kind of connection to a deeper, "noumenal" world.

Early Quakers used to refer to the "inward light," a concept that may have been lost for many modern Quakers and even for modern Catholics as well. I think this notion most clearly embodies a deeper sense in which spirituality, and perhaps also morality, is not dependent on self-consciousness but is something given to us by God. Nowadays people talk about the inner light and seem to think it's something that they've got, inherently and inescapably. But this isn't completely accurate, on my view. The inward light is much closer to what a Catholic would call grace, which is always treated as a divine gift. It doesn't come for free. It doesn't come with the equipment, as it were; it comes by the grace of God. You receive that noumenal access as a gift. It follows, if you take this language literally, that it would be up to God, not a finite designer, as to whether or not a particular constructed artifact would have this sort of access or awareness and hence whether it would have a spiritual life.

I have to admit that I don't find this way of speaking completely congenial. One of the ways in which I part company with my wife, and even with fairly thoughtful Catholic doctrine, is on the topic of providence, God's ability to act in the world. Physical intervention by God in the world, in any sort of direct sense, is not part of my conception of the spiritual dimension of things. I would qualify this statement, however, with the Quaker phrase "My hands are thy hands." We interpret the inner light not as God's intervening physically in the world, but as humans' seeking to do "his will" through that inner light. But my spiritual conception of things does not include God's intervening in the world, other than through grace.

Well, what about "his will"? It is a big problem for a lot of biologists, especially evolutionary biologists, to look at the world in terms of someone's will, that is, teleologically. I wonder how you would juggle these two different interpretations of how things operate in the world: on the one hand, looking at things purely mechanically, without any intentions and, on the other hand, looking for manifestations of God's will and, correspondingly, for human free will.

It's very tricky, that one, isn't it? In my opinion the field is littered with the corpses of theories that have attempted to derive free will from computational or physical properties. It's something that seems to be a disease that infects people at a certain stage in their careers. [The mathematician] Roger Penrose is the most recent famous example of someone from outside the field who basically says, "Oh, yes, we've got it. We can derive free will from quantum indeterminacy via computational modeling"—a claim that makes all the people in this [computer science] building, and people anywhere within a hundred meters of any of the centers of expertise in the area, fall down laughing. It's a "God of the Gaps" argument that has patent flaws. I doubt that such questions will ever be resolved starting from computational modeling in the attempt to derive conscious decision making.

What resources might a computational model lend to such a project, though—especially, say, in the case of decision-making processes?

I've put a good deal of thought into these questions. One first has to clarify what is meant by "decision making." Sometimes an external procedure exists by which it's agreed decisions will be made—whether it's a vote or a consensus or "the sense of the meeting," as we practice it in Quaker business meetings—but that's not enough to convey what we normally mean by "decision making." The phrase also has the connotation of making up one's mind. Yet this is one of those notions that, when you push hard on it, tends to run away from you; it's elusive, because it's hard to find any solid ground on which to define what decision making is.

For example, you might turn to computational systems as your model and say, "Well, okay, we computer scientists build behaving artifacts. We build things that people interact with and are prepared to impute rationality to—whether they are as simple as chess-playing robots or as complicated as English-language, interaction-based help systems for Microsoft Word. People ask such systems questions or pose problems to them, and they give answers. They exhibit behaviors and those behaviors represent choices. Hence they are decision-making systems. Not only that, but I can look at them and tell you in no uncertain terms exactly where those choices came from."

Yet there is a certain locatedness in computers that we can't impute to humans. Even if the computer generates a series of random decisions through a weighted coin flip, it lacks the intrinsic properties that most of us believe occur in decision making. In the end the computational model is not sufficient for understanding free will or what's going on when a human has an internal monologue of the sort that so often occurs: "Should I pick up that money and give it to that woman who's just dropped it? Or will I keep it for myself?" At some point, of course, you either run after her or you don't. But it's hard to find language for talking about what happens at the cusp of that decision. And I personally haven't found that knowing the closest analogs to that situation in a computational context is of any help. At some point the logic and semantics of the computational metaphor fall short of what actually occurs in simple human decision making.

But there is a tremendous allure to the computational metaphor, isn't there? Is it the potential for greater synonymy that drives people to look for computational models of human consciousness and even human spirituality?

I agree that that there's a tremendous allure there. But that drive can often be mistaken for arrogance. Why are we pedestrian cognitive scientists going to have any more luck than the greatest minds of the successive ages between, say, the early Greek philosophers and today have had at cracking these problems? There's been a singular lack of success for well over two thousand years, and the fundamental questions, both in understanding the human mind and in understanding human spirituality, remain. It's fair to ask, "Why should you cognitive scientists be any different? Why do you act as if the world has changed?" We tend to answer by saying, "We're standing on the shoulders of people who've gone before." A little bit has been learned over time, so we can do a little better.

But there is also the belief that we have a qualitatively different intellectual tool in the form of that computational metaphor and its realization in real physical silicon and metal and plastic. The ability not only to use the computational metaphor as a theoretical tool but also to construct computational artifacts whose behavior we can then tear apart in a way we can't do with human beings, right down to the lowest level if we need to, does make a potential difference. Still, I think the metaphor, and its attendant artifacts, have led to much more success in the semantic, mental, and psychological domains than they have in the spiritual or theological domain. For somewhat obvious reasons the computational model has had considerably more efficacy for talking about the construction and movement of language than it has for understanding the formation of belief, especially belief in God, or even for understanding conscious decision making and free will. But there might be some leverage in the domains of language and psychology for accessing the spiritual.

And what might that leverage be? Isn't spirituality of such a different type that it might make "leveraging" impossible?

What has often occurred is that gains in psychology or linguistics have been used as a foothold for exploring other areas of consciousness and mental processes. But, possibly for the reason you raise, I'm not aware of any examples where scientists have leveraged themselves into the spiritual or theological realm of human consciousness in this way.

Fascinating. Part of it, and I guess we touched on this before, may be that it's so difficult to say what spiritual life looks like. It's not identical to the ethical life; that's relatively easy to describe. But the spiritual life is basically invisible.

And those who model the spiritual life are likely to have very idiosyncratic views on it. For example, how does one even intelligibly—and, much less, scientifically—analyze the mystical perspective, which essentially states, "I know that God is greater than I know him to be"? How does one talk about a negative theology in scientific or computational terms? Any attempt to frame propositions about this form of belief or way of life is by definition inadequate and misleading. That doesn't leave much room for scientific inquiry. And I'm not sure whether the spiritual life—as mystics describe it, for example—has any analogs with the other aspects of consciousness that scientists have been able to model, such as linguistics.

What about God's will on a day-to-day basis? Is that a part of your world?

Absolutely. It's crucial. It's because the Quaker business meeting, the way we run our meetings as an institution, models the way we run our lives. We say very explicitly, "We worship together with a concern for business." We take the procedure of the Friends' meeting for worship as the mechanism by which we do business. What that means is that in serious issues we are seeking God's will for the meeting, even in instances where we're doing business. We are not trying to achieve consensus. Consensus is a secular process. The meeting for worship for business works when there is a sufficient number of people who see themselves as participating in God's will for that meeting. This doesn't apply to everything, of course. As somebody put it not very long ago, when what you are talking about is whether to have chairs or benches in the meeting room, it is unlikely that God has an opinion. Also, people mean different things by the phrase. The best you can hope for is something that will be pleasing in the sight of God.

"The will of God" is of course code, because we are stuck using these words that project a humanoid perspective onto the referent of that word *God*. Language is never sufficient, especially in the instance of God. But what doing God's will means for me, and I think for many more intellectual Quakers, is acting in a way that is consistent with bringing about the kingdom of God on Earth. There is an incredible quotation from George Fox, which says that the Second Coming has already happened: "Christ has come to teach his people himself." You can receive Christ's guidance—in the gathered meeting or even on your own. He promised it: the Paraclete will come, "I will be there." That is very much God-talk, very much Christ-talk, which in one way I find somewhat off-putting and unhelpful. On the other hand, though, maybe it's really real. Maybe what the language means is that the ineffable, the noumenal, the abstract ground, God, is accessible to all of us. If we can get around the various problems of being misled, and misleading oneself, we may have access to it, since it is universal.

And this would imply that the kingdom of God simply means manifesting this ever-present will of God, right?

Yes. Indeed, I would agree that the notion of "manifesting" is right, because it's not simply a matter of belief. The Quaker always asks, "Of these behaviors, which is the one that is most consonant with what the noumenal essence would have? What would it mean to have the kingdom of God on Earth?" Doing God's will means acting in as consonant a way with the kingdom of God on Earth as we can. And "the kingdom of God on Earth" means, in non-God-talk, living the life we are meant to live, the life that is the best expression of our humanity. At the end of the day that's the answer for me, and for a lot of Quakers, to the question: How ought I to live my life?

How does that notion of consonance with God's will, or consonance with the project of bringing about the kingdom of God on Earth, mesh with the idea that's so popular among evolutionary biologists these days: that life is the product of the repeated application of algorithms, that we're not moving from one place toward another, that heaven on Earth is something that will be defined by the course of evolution?

First of all, as a linguist, I quickly learned to think in nonteleological terms. One of the things we've learned in the twentieth century about human languages is that they change and they evolve, but they aren't heading in a particular direction. Languages move around. English has gone from one end to the other of various possible spectra, such as whether a verb comes in the middle or end of a sentence. We've changed our minds about that matter three or four times in the recorded history of English and its predecessors; there isn't any right answer. It's not going to be the case that five thousand years from now all languages will be verb final or all languages will be verb middle. Languages aren't going anywhere, they're just going. I have no difficulty with the notion that there are lots of evolutionary systems, in the various senses of the word, for which the same thing is true.

But it seems to me that the progress toward the kingdom is orthogonal to all that. For example, in Hasidic Judaism there are, it turns out, six hundred-and-some-odd clauses in the law. And there is a fairly eschatological hope that if everyone were to observe all these laws for just one day, then that would bring about the kingdom of heaven on Earth, or at least the necessary conditions for it. Of course, that by no means implies it's going to happen or that history is leading inevitably toward such a moment. It is tantamount to the possibility that all the air in this room will coalesce in the upper lefthand corner of the room long enough for me to choke. You can measure the probability that that might happen. It just needs all the right conditions to come together at the right time. There's no progress toward that event; it just may happen one day. You could take the same perspective about fulfilling the Jewish law [the laws in the Torah, the five books of Moses]: it might happen one day, but there is no historical teleology [that is] moving history in that direction. It remains a possibility—a beautiful one, however unlikely.

But isn't the law like languages in that it too is always changing, always evolving?

That's right; history brings constant change. So it's incumbent upon individuals to understand for themselves, as best they can, what "living the kingdom" means and then to try to live that way. Because God is transcendent, and because we are always located in our own situation, our human access to God is inevitably limited. But we do have access; that's my article of faith, I suppose.

In light of all the different religions, one might ask, "How do you know whether your claim to have gotten hold of a bit of the transcendent is accurate?" That's the question of authority. Here's one of the respects in which Catholics and Quakers are more alike than many of the other Protestant sects. For neither Catholics nor Quakers is the answer, "Oh, it's in the Bible." For both of them—more or less, depending on exactly what kind of Catholic you are—the answer is, "through the tradition of the church." We work it out together. The problem of authority is worked out through a historical and communal effort. In

Quakerism there is no equivalent role to the pope, but there is a surprising amount of consensus that it is the "consent of the faithful," to use the Catholic phrase, that determines what the transcendent is telling us. What underlies this consensus is the belief that there is a consistent story that underlies history and that we can get access to it. Imperfectly. From time to time.

12. Robert Pollack

Dr. Robert Pollack is professor of biological sciences, adjunct professor of religion, and lecturer in psychiatry at the Center for Psychoanalytic Training and Research; he is also director of the Earth Institute's Center for the Study of Science and Religion, all at Columbia University. Pollack graduated from Columbia College with a major in physics. He holds a Ph.D. in biology from Brandeis University and has been a research scientist at the Weizmann Institute and at Cold Spring Harbor Laboratory, an assistant professor of pathology at New York University Medical Center and an associate professor of microbiology at the State University of New York at Stony Brook. Pollack has also been a professor of biological sciences at

---

Philip Clayton interviewed Robert Pollack; Kevin Lucid edited the interview for publication.

Columbia since 1978 and was dean of Columbia College from 1982 to 1989. He received the Alexander Hamilton Medal from Columbia University and has held a Guggenheim Fellowship. Since 1997, he has been a member of the Century Association, a private club in New York City that was founded by William Cullen Bryant and named for its one hundred members. He was also president of the Hillel of Columbia University and Barnard College from 1998 to 2001.

Pollack is the author of more than one hundred research papers on the oncogenic phenotype of mammalian cells in culture. His 1994 book, *Sign of Life: The Language and Meaning of DNA,* received the Lionel Trilling Award. Pollack's latest work, *The Faith of Biology and the Biology of Faith: Order, Meaning and Free Will in Modern Science,* was published by Columbia University Press in 2000. He is a Fellow of the American Association for the Advancement of Science and the World Economic Forum in Davos, Switzerland; a member of the American Psychoanalytical Association; a former director of Nutrition 21, Inc., a nutritional bioscience company; and a member of the scientific advisory board for Tapestry Pharmaceuticals, Inc., a pharmaceutical company that specializes in developing cancer treatments and therapies from "natural product sources."

Robert Pollack is an intense discussion partner. He's ready to go much deeper with a question or topic than one would expect, and his responses do not shy away from even the most controversial topics. Nor is the critical probing limited to people who are not present: the questioner can quickly find himself under question as well. Pollack is driven by uncompromising determination to say what is true and to know what is right—exactly the twofold position that he describes in the interview.

In your career—from your early physics studies, through your work in cellular biology, and now on to your recent series of books—you seem to have moved from questions on a very small scale to questions of broader and broader significance. How would you characterize the growth of your career as a scientist?

I think at a very early age—way before college and probably sometime as a kid—I was made aware that something existed in the outside world that could be understood by quantitative measurement. That concept developed into something fruitful for me. As a scientist I scope out the possible quantitative ways to test ideas. So it has been in my work in physics, in looking at oncogenic viruses in developmental biology, and in writing books about science.

Someone once said, though, that life is lived forward and understood backward. What guided my career forward? I think early on in my undergraduate work at Columbia University, during a series of discoveries made with my fellow graduate student Arno Penzias, I saw and experienced the *discontinuity* between what science in fact offered me and what the content of that emotional experience of discovery told me. I came to this understanding emotionally as opposed to articulately, however.

I was drawn to science because I wanted to be part of that experience of discovery. I felt connected to it. I wanted to learn more about what it meant for me. I got a kick out reading about it, and I get a kick out of sharing and teaching the experience to students. I'm a good teacher of undergraduate biology and graduate biology because I make sure the students find that sense of discovery for themselves by reading original papers by the scientists who experienced the discoveries firsthand.

I also think I am entitled as an observant Jew to say what my observance teaches me about the boundaries of acceptable behavior within science, at least the science I know best. It teaches that certain behaviors in science are ethically better than other ones. These boundaries of respect and responsibility for someone else's free will within scientific practice are more important than my ability to tell somebody what to do. Both qualities appear throughout my scientific work.

When I read your descriptions of how our world is mathematically expressible, they remind me more of the typical attitude and enthusiasm of a physicist. What brought you to doctoral work in biology?

Mine is a story that cannot happen today because of the privatization, commercialization, and professionalization of molecular biology.

At the time I received a B.A. from Columbia, I had a spotty academic record, but I knew that I was bright and that I talk and write well. But I knew I had no career. I took the GREs in physics, but the test did not go spectacularly well: I scored 97 percent or so—not enough to get a fellowship at Columbia. I took that as a sign that I shouldn't do physics. So what should I do?

I read, and afterward, I found I should look into biology. I went to Brandeis to be interviewed in the physics department, and I found that they had a molecular biology program run by a biologist who was trained in physics. When I met with the professor at the physics department, he walked my application one floor down to the biology department and admitted me on an NIH [National Institutes of Health] fellowship as a biology graduate, along with enough money to get married.

[Laughs] No, that certainly wouldn't happen today. But the professor at Brandeis recognized your potential?

Yes. As an undergraduate at Columbia in the late 1950s, I was at a place that had more Nobel laureates who were going to come out of that lab run by Isadore Rabi than any other physics department anywhere in the world. Eleven Nobel Prize winners came out of that lab. That's off-the-scale different from the mean. But I can't claim any great insight about why he recognized my potential over others' beyond the luck of being surrounded by such terrific people at that time and taking it for granted that those high standards were what I should demand from my approach to science.

Those kinds of people at the lab saw that there were ways to approach biology through physics. They saw not *biophysics,* as such, but that physics principle of looking for the simplest possible system. The approach required stripping away all variables that you couldn't control and matching the simplest building block with the minimal definition of being alive. That is what they would call biology. They didn't postulate the wedding of chemistry to biology but the wedding of the physicist's idea of a simple, elegant, quantitative model with biology. This imprint was what I took to Brandeis.

What did it mean, more precisely, to take this background in physics and apply it to molecular biology?

The real questions for me came when I removed the thought of the luck that happened at that particular age and with the cohorts I kept. I was led to find that the ancestry of these notions was in Mendel's idea of genetics. When you read Mendel's paper, which is nearly 150 years old, what jumps out at you is the abstraction of it—the complete absence of a biological entity to explain the behavior of the characters that we now call genes. He says at the end of his paper, I don't understand how this can be, but this is what I see. To my mind Mendel was in the world of physics at that time. He was a statistician so he applied statistical principles to his counting. But the idea that you reduce a biological question to a quantitative one by stripping away the variations worked against the entire spirit of biology, even fifty or one hundred years ago. At those times biology was meant to understand the *diversity* of nature, not the *variations*.

You were already familiar with that discontinuity of Mendel's emotion with his discoveries. So what you are saying is that molecular biology as we now know it could not yet flourish at that time, even given Mendel's luck at his particular age. But the foundations of molecular biology were built on these successive discoveries.

Absolutely. Molecular biology was invented by physicists who became biologists. Max Delbrook is the father of molecular biology. But I think Irwin Schroedinger wrote the most prescient book there is in science called *What Is Life?* It describes what DNA must be on the general physical principles of what inheritance tells you it must be. You read that book and you say, "This is uncanny. This guy must have known where science was about to go." It was another ten years before that book formed in the minds and in the approach of people like [James] Watson and [Francis] Crick. In the laboratory of a physics place—and, remember, crystallography is a physics place—Watson and Crick asked the right question: How can a structure of a molecule carry genetic information? I consider that to be the question of the century. It was answered by physicists applying the principles and tools of physics to a biological question. I'm in that school.

And Crick and Watson's results were just being digested as you were completing your physics training and beginning Ph.D. work. What did the NIH grant allow you to do?

I studied something in phage genetics and went from there to study cancer as a disease, reducing it to a question of the behavior of a virus so small it has only four or five genes, and yet it has the capacity to convert a normal cell to a cancer cell. This was my first real science: a study of cancerous cells where I basically applied a piece of physics thinking onto biology through this medical research. The approach paid off. I really made a bump in the medical field.

What I did was to take tumors and grow them in culture. I'd take a single cell and I'd grow from that cell a clone of tumor cells, so that is as genetically homogeneous as you could get in the system of mammalian cells at that time. Then I'd let these cells reach a cell density in a dish that emulates tissue density. Normal cells should stop dividing under that tissue density. That's what a normal tissue is. It's a bunch of cells that are under growth regulation by density.

That was a remarkable discovery, but what led you to discover this?

I intuited that if you apply what is called, in bacterial genetics, negative selection and kill the dividing cells under that condition, any preexistent Darwinian random variants, which are more normal than the tumor clone should be, will survive the killing. I then selected revertant-phenotype, normal-behaving cells that still contained within them the viral genes responsible for the initial transformation. In doing this, I showed that the virus is working through modulation of cellular genes, and if you knock out one of those cellular genes, you get the normal phenotype back.

What did this mean? In medical terms I did something that I think is as important as anything I will ever do in science. I showed that, intrinsically, the growth of a tumor in a body is not an irreversible fate. What it represents is Darwinian selection for the worst kind of cell against the body's defenses. If you rejigger the selection system so that the better kind of cell is the only kind to survive, the cells that vary in the direction of normality will survive. This means that what is present, always, is intrinsic, random variation. Today we know that once p53 deletion ups the frequency of surviving mutant cells, all kinds of variants will emerge that can survive any chemotherapy.

Is this when your values started to figure in your science ... when you saw that your approach improved medical understanding?

I did not come to value the relation of the two from that instance; my values adapted to the situation and existed before that work. There was a moment, however, when I became quite conscious of how important my values became in my choices at work. In my field the way investment money came to us was changed by the Bayh-Dole Amendment. The amendment obliged the NIH to seek private funding [through exclusive licensing] for the discoveries made in NIH-funded laboratories. At the same time the Hughes Institute came in with separate money for top labs. So the notion of being a government servant and looking at nature in order to help people was diluted by money coming from these well-meaning funders and by the push for profit from the government. My values to act with autonomy and independence took hold as I saw this happening. I became opposed to institutionalization, corporatization, privatization, and working in big systems, but I had become obliged to play a game that was as fraught as Defense Department–supported physics. I got away from that game by becoming a biologist.

I remember at Columbia, there was a guy named Walter Faust, a graduate student in the same lab. Faust received a $50,000 grant from the Defense Department because his work was on what we now call lasers. At that time lasers were only just being discovered. His work involved developing a laser that worked in the infrared spectrum. The Defense Department wanted it so they could detect the tails of rockets being fired from overseas. Faust hired technicians and got his Ph.D. a few years after he started on that grant. When I saw this, I thought, "No harm to him but harm to the institution." Columbia was now assisting the Defense Department to achieve its military ends.

I'm seeing two things here: an intellectual movement toward these more holistic questions from your early training in physics but also a certain growing skepticism about the politics of science, about science as an institution.

The intellectual movement you pegged exactly right. The political movement, socially and sociologically, was actually a movement around me, while I stayed still. I don't think I changed in terms of what I think science is about. I think that the science I first entered into changed to an institutionalized place where I didn't choose to go. I write that in my books over and over again.

There is something different in the Signs of Life book than the way you described your entry into biology. Can you describe what changed in your view?

These are tough questions for me because I am tempted in two opposite directions. I'm tempted on the one hand to tell a pretty story that makes sense of the stages of my career, but I'm also tempted to be honest about data—which is to say, I know none of these life decisions were rational at the moment they happened. All throughout my career I was like the roadrunner in the *Road Runner* cartoon. I ran off the edge of the cliff, not just away from the wolves, but because I didn't look down, I just kept running.

I went from Charles Townes's laboratory, [where I was] working for Arno Penzias on the construction of instrumentation, which led to the discovery of the big-bang data point. I then entered a postdoc program at the NYU pathology department, where I was able to apply the principles of molecular biology to cancer, helping to bring cancer into the world of molecular biology with great fruitfulness.

Then I started work at Cold Spring Harbor Laboratory. There I reported directly to Jim Watson and was member of a group of guys in their early thirties who were setting up laboratories with NIH funding and private funding. The laboratory was a pure situation, the purest kind of science I can imagine. Watson stood between us and the world. He was completely altruistic about not putting his name on anything we did, and also we had no tenure concerns, no teaching concerns, no money concerns, no calendar concerns, no private concerns. Two-thirds of the marriages broke up there in five years. I had to leave because I valued my wife and my child more than I valued my work.

You left your work at the Cold Spring Harbor lab because that environment was too intense?

Too intense, yes, but, more important, I found myself in an environment where nothing else seemed to matter to my colleagues. If somebody's wife developed cancer, as happened to a colleague of mine, I was told that Watson paid for the doctor to send the woman back to Germany so that the guy wouldn't be distracted. That's the kind of stuff I am talking about. The best example I have that mimics my experience at Cold Spring Harbor is Solzhenitsyn's book *The First Circle.* It is a brilliant novel of an imagined situation in which Stalin decides that it's not enough to wiretap phones because everybody knows the phones are wiretapped. He wanted to know who is the person making the call. So Stalin decides he needs quantitative, computer-driven voice recognition instead. He arrests all the appropriate people to develop the technology of computer voice recognition. He arrests them on what he claims are political charges. Only the jail is a laboratory in Moscow. He furnishes them with all the Western equipment they call for, but they cannot leave. They are driven around in bread trucks. The point of the book is—they love it.

That is what Cold Spring Harbor was. Nobody said you couldn't leave, but there was such a strong selection for work that people were becoming dysfunctional for the world. You didn't have to carry money. In religious terms I was not an observant person at all then, but it was amazing to me that nobody knew what day it was. The notion of the sabbath was perpendicular to the operation of the place. The atmosphere created and nurtured the perspective that where you are didn't matter, because where you are is determined by where you are in your work. It reified work to God. It made work the reason for existence, and it provided the resources to make that work. I have never been, nor can imagine being, in a place where, for no other reason than a good idea, you could have all the resources possible to test your idea.

What effect did the experience at Cold Spring Harbor have on your attitude toward your scientific work?

At Cold Spring Harbor I had found the best position for doing science that existed in the world at that time—but I couldn't work under those conditions. Not that I couldn't do the science. I pumped out more papers there than before or afterward, but I couldn't enjoy it. I didn't have the emotional kick I had when I approached science as a kid or [was] working for Penzias or doing my work in a pathology department. That fabricated world wasn't connected to life and death. It was emotionally blinded.

But as a scientist—let me put it as strongly as I can—my work and the others' work at Cold Spring Harbor denied data. As pure as it was about the data of the laboratory, it was fraudulent about the data of the heart. You could not be a full person there. There a scientist could not say, "I'm scared of sickness. I want to work on this sickness because it troubles me." You could not say, "My father got Alzheimer's. I want to worry about that and take it on for that reason."

But, sadly, isn't that how great research comes out of these fields?

Yes, so I felt. I don't want to say this as if I came to this insight only by my autonomous nature, though. I've been married now almost forty years. My wife is an artist; she is not a scientist. She reads people by her perceptive assessment of their character, behavior, and presence. When she saw, she warned me—alerted me—over and over again that this situation was not healthy. It was not a healthy situation for *us*. Our life together was predicated on an honest assessment of our feelings as well as our intellectual capacities, and we recognized that our feelings were being submerged, denied, ignored, obliterated, suppressed. I was forced to accept my inability to become the person I had thought I had wanted to be. I would say that is—forgive me for a jump out of the frame, but in retrospect I would say—that was my first religious experience that I could acknowledge.

But that's not a jump out of the frame at all. The story that moves from biological research to the concern with science and religion is the story from the very specific to the holistic. It's fascinating that the time when you turn away from the mechanistic, physics-based approach to biological science is also the time when the religion begins to become important.

I think that the causality, if it's there in any direction, is backward. I first became bereft of connection to other human beings and from that decided to leave the work. It wasn't that the work drove me away. The absence of emotional bonding, the absence of real, true, deep, reliable human connection—a failure to recognize and value those connections above all things—came to me packed with great force because, by imitating that model in those experiences, I was in danger of alienating my own family structure. I saw what was really important. I asked myself, "What do I want to base my life on?" The simple answer would be, "Don't be a jerk. You can have both. You have a private life too. This is America." But the fact is, America or not America, science as a religion is too demanding. I would say in retrospect that I experienced the invitation to join the high church of science and found I was not a member of it. I could imitate belief, as I'm sure in every religion there are people who go through the motions, but I'm not that kind of guy.

The demands of the scientific priesthood were too inhuman.

Too inhuman. I couldn't do it. We reached a point in our lives where our daughter was getting close to thirteen, and we had to decide, are we Jewish in any way besides our parents' being Jewish or not? If we are, then we had to break completely away from the mocking tone that science often takes about any religious observance and do something in our private life that would be outside that frame. And we did. Having done it, I realize that this is as important to me as anything I'm doing in science, and I'd better damn well pay attention to that fact.

Was the beginning of serious observance in conflict with science as such, or did you experience it as supplementing the "science side" of being a person?

I see the categories you are using, but I don't like either of them. Observance was neither of those. Observance was in service to feelings. Now, there are people, and among them very good people, for whom feelings are served through science. I'm not one of those people. There are great scientists who are humane characters and who have no religious conviction that I've ever been able to surface, but I haven't the strength to be one of those people. I wish to be very plain about that. One of the people I am grateful to have as a teacher is a rabbi named Adin Steinsaltz. He is a translator of Talmud into many languages. He lives in Israel. I refer to him in my book, in *The Faith of Biology and the Biology of Faith.* I asked Rabbi Steinsaltz the following question: I had been confronted by a colleague who was a very articulate nonreligious person. A person whose articulate nonreligiousness stems from his certainty that there are no data for the presence of any governing power beyond the data. To me it's a circular argument of saying that the absence of data is data for absence. Our exchange had challenged me with the realization that—point for point, issue for issue, behavior for behavior, value for value—this man and I are identical in our ideas, and yet he gets there without God. Therefore I had to ask myself, "What am I talking about?"

That is a powerful critique. How did you respond?

I was unable to answer it. I went to my rabbi and asked, "What do you say to this?" His answer was immediate. He said, "There are people for whom the throne of God is empty. When you find such a person, you hold tightly to them because they are among the most powerful and good people in the world. But most people who claim the throne of God is empty actually secretly put themselves on it, their money on it, their boss on it, their work on it, some hoist a flag onto it. At that point they are doomed to idolatry, and you stay away." And I thought, "That's great." From there, what he said to me about himself, and me, was true. Those of us who accept God on the throne of God are too weak or too small, or too bounded in our immodesty, to claim that anybody or anything else should be on that throne—and we can't live with an empty throne.

I think in the end it comes [down] to an emotional state. That is why I am so strongly emphasizing emotional rather than intellectual judgment. Emotionally, to say that there are no data for the presence of any governing power is too empty. Too empty, if you don't put *yourself* on that throne instead, but you simply say, as Sherry [Sherwin B.] Nuland says in his book on dying, "It's great to die." All my big molecules go back to small molecules and get incorporated into other living things, and what is wrong with that? If you can pull that off, it's terrific. I can't do that.

The trick to being a scientist is when you can't do something, you have to say so. You can't just say what people want because then it makes them nervous to tell the truth. You've got to tell the truth. I realized that my truth is, I can't do it. I can't live in a world in which that throne is empty. And I will not put myself on it because it is in such bad taste to do so, even if many scientists do. Some put science itself on that throne; the future capacity of science to discover anything it wants then becomes the occupant of the throne of God. That is as if to say that some number of new facts will provide the reason for living. I can't put science on the throne; I don't think I ever tried to.

I didn't see the bigger question clearly until my kid got to a certain age, until I saw my parents reach a certain state, and until I realized that the immortality of discovery was an all-too-fleeting version of the immortality that people think about. More data was not enough for me.

Many people share with you this critique of science, of science's being placed on the throne, and then conclude, "Science and religion stand in a fundamental tension." Do you agree with this sense of an ultimate incompatibility? Or is it possible, once this insight has been reached—once one becomes observant or finds a religious faith—that one can return to science with a different attitude?

I have a two-part answer for that one. First the argument, then the feelings. The argument makes sense to me, but it's a controversial leap. Science makes the following claim for itself, legitimately: most of what is knowable is unknown at this moment, and most of what is unknown will be knowable eventually through science. The *faith* of science makes a further claim: *all* that is unknown will be knowable through science. The distinction between the two turns on the question: Is there anything unknown now, whether or not it lies on the outer edge of what is knowable, that will *never* be understood, anything that is ultimately unknowable? No one denies that science will push the margin ever *closer* to full knowledge. The issue is whether some unknown will always remain. That question about science is by its very nature not answerable by science. Therefore to claim there is nothing unknowable is *an act of faith,* and to affirm this statement makes science into a faith.

Further, this statement is incompatible with any other religion, and it's this statement of faith that makes it incompatible. The actions of a scientist, short of that statement of faith, are completely compatible with any religion. The actions are to understand God's creation. But to say further that the understanding reached is the only understanding *possible* is neither right nor wrong. It's untestable and therefore unassertable by an honest scientist within science. It is only sayable by a person of faith within faith.

A person who puts all faith claims in science as the locus of faith cannot then also say, as another statement of faith, God sits on the throne and therefore the universe is reasonable as well as understandable. That's my distinction. It depends, in other words, on which scientist you are talking to. There are some scientists who, regardless of whether they have any faith or not, do not have faith in science as a religion. There are others for whom the faith in science as a religion makes them zealots for science and angry at any other religion. And in fact it makes them *missionaries* of science. Like other missionaries, some would kill you rather than lose you by conversion.

Some very popular, widely published people seem to take that attitude.

I don't think that the missionary zeal emerges from science itself; it emerges instead from a religion that cannot be part of science because it cannot be tested as a hypothesis. This is the important thing to realize: science lives in the world of testable hypotheses. The hypothesis that there is something unknowable to science is not testable. You either say this or you don't, but what you say is as much a statement of faith as the statement of any religion.

Let's explore the other paradigm. Describe for me what the practice of science might look like for people whose religious faith is classically religious—Jewish, Christian, or another specific religion—and who continue to practice science.

Here I can only mention *The Faith of Biology and the Biology of Faith*, published by Columbia University Press, since it is my answer to that question, my attempt to say what I think my science should look like from the point of view of my religion. The answer is intrinsically and intentionally personal. I don't think it is right as a person of faith to say what a Christian should do with his or her science. But I think I am entitled as an observant Jew at this point to say, "This is what my observance teaches me about the boundaries of behavior within the science I know best." What it teaches it that there is a world of important, critical medical science to be done now, first rather than later, because of religious convictions.

My answer in the broad sense is that one should be able to say, "My religious faith teaches me that these questions in science are more important than those questions." These behaviors in science are better than those behaviors. These boundaries of respect and responsibility for someone else's free will are more important than my ability to tell somebody what to do. These social structures protect religious faith in the patient, the doctor, the scientist, the administrator, and the dean, whereas social structures like those can make everybody slaves, de facto—the way Jews were slaves in Egypt. I therefore think that the first answer is that a religious person who is a scientist must come out from behind the bench and enter the world. You should not take positions that are called political just because you are religious but because you are also a scientist. You must care about the world and not just your work. All of those are answers. Does that make sense?

That's a very clear answer. Many of the implications you just hinted at are ones that you explore more fully in your book. Perhaps the most controversial implication of your approach to integrating scientific work and religious practice is that one's religious commitment would affect even the types of questions that one asks and researches. Since that's the claim that, for a lot of practicing scientists, will be the biggest shock, could you describe a bit more about what it entails?

We could light a bomb here and talk about abortion. But let's set that aside. Not because we should never talk about it but because it is too explosive an issue at the moment and thereby prevents further discussion. Instead, of the two universal events around which we revolve, birth and death, let us examine what death is and the distinction between dying and being dead.

I would say I am taught from my religious tradition, written and oral, through prayer and through study of texts, that I am obliged—for people in my family, for friends, for strangers, for the religious and scientific and medical structures I am a part of—to assure that a person until dead is socialized in the world to the extent they are capable of being so and never [is] abandoned on the grounds that there is nothing more to keep them alive with.

To be treated as human beings in the fullest possible sense all the way through their lives—

To the last moment. The Talmud is stunning about this. A person who is defined by the best medical ability of the day as certain to die within three days may marry, may sign a contract that is legally binding, may vote, may divorce his wife, may do anything that a man may do otherwise. By the same token, a person who is going on the sabbath to pray walks by an accident, a building falls down, and he sees somebody is going to die from being crushed within the next ten minutes. He is obliged not to pray but rather is obliged to save that person's life. As a religious obligation he must "violate" the sabbath to save a life.

Now the third story from the tradition. Someone wishes to say the final confession and die. He is tired of living, he is suffering, and he can't concentrate because a woodcutter is chopping wood outside his window. You are allowed to tell the woodcutter to stop chopping. And by the extension of this tradition to medical practice, you are allowed to withhold interventions that keep a suffering person alive.

I am the most trivial of Talmud scholars. At this point I depend entirely on friends to point me in the right direction, although I'm starting to study a little more. But I see how rich this tradition is in providing guidance to me and my colleagues. When I teach people who are going to be doctors, or when I sit on panels of the medical school curriculum committee, as I do, and when I write books, I try to steer the ship a little bit away from the machinery that puts you in a room with tubes and coldness, and you die only connected to machines. One doesn't have to do that. I admire the Anglican and Catholic hospice movements from the bottom of my heart, and I work, as a scientist, to make that kind of facility available to people of my own religion.

What I'm understanding is that one's religious commitment, and the values of one's religious tradition—yours and others'—lead to the humanization of the medical sciences.

Correct. I would say, again at a deep level, that such an approach is absolutely legitimate and not just fancy talking. Because at the deep level, having had the gift to experience it a few times, I recognize an astonishing similarity between the moment of scientific discovery and the feeling of God's presence—a revelatory state. Both of these moments are very rare, very transient. You would be lucky if the experience happens even once. But when it happens, what is common to both moments is that you are overtaken. You are not in charge. That should lead to modesty in science, as it certainly leads to modesty in religion. I think the convergence of behavior in science and behavior in religion is not driven by some kind of synthetic religion that includes science but is driven by the similarity of the data. A human being can be overtaken by insight, and a human being can be overtaken by revelation. Should a human being be lucky enough to have that happen in both ways, that person ought to have the wit to see that the data tell him or her to be modest and to care for others.

In theological terms is this not a way of saying that there are two forms of the manifestation of God? There can be a more direct revelation, say, through religious experience. In your tradition is there also an indirect revelation of God's nature in the physical world?

I love the question, and I'm going to flip it back to you by answering it with the full ambiguity it deserves. That is to say, you asked me about my tradition. But we are talking about two different traditions. In the tradition of science you would not state the question that way. You would say it something like this: there is a brain state, a neurological organization, driven by the history of the organism and by the sequences of DNA that constructed a brain that has the capacity to experience such brain states. We also recognize a brain state called insight that comes from revelation. Here is the point: the same brain state, whether it comes from one place or the other, is common to both traditions. Eventually, science may understand that brain state. I would say that there is an example of a piece of science and [of] religion's coming together. What is that brain state? What happens? What's going on in the brain of a prophet that doesn't go on in the brain of you or me? What goes on in the brain of a genius that doesn't go on in the brain of you or me? Where does the difference lie? It doesn't lie in a DNA sequence. It lies in the sum of a DNA-constructed brain and a life's experience. Where is it? What is it? Those, I think, are questions that are wide open at this point. And yet one can restate the overlap or similarity of the two experiences without mentioning God.

Now the religious perspective. I'm sensitive to the fact that when I'm talking about religion, I'm really talking about Abrahamic religions. I have a very wonderful Buddhist colleague, Bob Thurman at Columbia, for whom a lot of this is nonsense. When you deal, as he does, with reincarnation and the absence of a Creator God, then a lot of what we talk about doesn't fit his religion at all—and yet he is very religious. I really struggle to find inclusive explanations, but I'll fall back on the words I am comfortable with. So take the example we discussed earlier. When your faith is that God sits on the throne of heaven, then you might say this of the larger question: In creating the world God says, as it were, "Give me a creature who can choose to pray to me or not, and then I will watch to see if the creature chooses to do so. But I don't care how you get that creature." In this context, for humans the purpose of everything is to ask questions like this, to be in conflict like this, to wonder about this, to struggle with this—and to decide in the end to accept it. To worry in the end that you made the wrong decision and then go back on it. All those struggles and affirmations legitimately become part of God's intention. Otherwise, why not be created as a machine that can just run well? God's risk in creating free-thinking creatures who can choose to do evil is an astonishing statement about the natural world.

That's how I would answer your question. The scientific insight—to discover an aspect of the natural world that can then be made useful—and the revelation—to discover that God's intention is that we live this way and not that way—come to the same thing. They come to a piece of God's intention: that we choose to figure out what to do and then do it.

Some people take this interconnection of the two spheres of science and religion so far as to say that the full substance of God's revelation can be revealed by science or, coming from the other direction, that religion can judge the results of science. For instance, some argue, "I know that Darwinian evolution is wrong because my inspired documents show that the world was not created that long ago." Do you agree? Or is there a certain autonomy of the two sides, a point where the overlap ends?

I think it's wonderful that we got this far without that question's arising, since most discussions of science and religion begin and end with that question. So let me make a dogmatic set of statements. First, neither I personally, nor the Center for Science and Religion at Columbia, have any intellectual or emotional investment in the idea that our purpose should be to find God's thumbprint in the natural world. Nor do we have any idea or intention to naturalize religious observance, to show that, for instance, *kashruth* [keeping kosher] exists in the Jewish tradition because it is hygienic.

What is demanded is demanded, and what is understood by data is understood by data; in that sense there are nonoverlapping aspects of human behavior. But where science and religion merge is in how you or I as individuals deal with the simple requirement that we must make a decision on a matter of faith. Do you or do you not think that the sum of what will be understood by human beings is all that is? That cannot be a statement of data; it must be a statement of faith.

So, again, I don't think that the categories, the "magisteria," are science and religion. The magisteria are of those who believe everything is understandable and of those who believe some things are intrinsically not understandable. You will find scientists and religious people in both categories. I think the world of scientists and religious people who accept that not everything will be understood is a world in which people can talk to each other with complete ease and naturalness. One by revelation, the other by data.

The world in which people say, "What is known now or will be known by human beings is all there is," is a world of rigid fundamentalism, both in religion and in science. That is, as I explained, not a world that I like to be in because there is nothing to say. The worst outcome in the world, I think, is that somebody turns up a proof and says, "Here are the ridges of God's thumbprint on this rock." What do you say? You say, "Then there is no work left. There is nothing to do anymore. There is no struggle. There is no gift to God in prayer anymore because it is given."

And no faith.

And there is no faith. Bob Thurman told me the following story. He is the Dalai Lama's representative in the United States and was once with the Dalai Lama while he was being interviewed by Carl Sagan on television. Sagan said to the Dalai Lama, "Your Holiness, what would you do if you saw the evidence from a research project that proved to you beyond doubt that reincarnation was impossible on the grounds of the second law of thermodynamics, the return of complex molecules to smaller ones, breaking the information that is present in the larger ones, the facts of universal entropy—" and on he went. "And you became convinced that reincarnation was not possible. What would you then do? What would happen to your faith?"

The Dalai Lama said, "Well, I would stop believing in reincarnation, of course. I'm not stupid. But tell me, do you know of such an experiment?" That was a great answer, and Sagan basically ended the discussion. There was nothing he could do. Sagan had hoped to get him to say, "I am a fundamentalist. I turn away from evidence because my faith is so strong." I agree with the Dalai Lama 100 percent. God's creation is to be understood. Beyond what is understandable is God. The creation is not the Creator. People who mistake the creation for the Creator are in the end worshipping idols.

But the devil sometimes lies in the details, especially in the fields in which you work in the medical sciences. There can be guidance from the religious traditions to make sure that death is not dehumanizing. There also can be guidance from them to determine what kinds of questions are important. But there is a fine line between the guidance about what questions can be asked and about what answers are allowable. It's in those gray areas that the difficult question of overlap arises.

Okay. In my world there is rhetoric and there is reality. The issue is not "where does religion affect the rhetoric?" but "where does religion affect the reality?" The rhetoric goes like this: All scientists are members of the guild of free men and women. They each make decisions as to their work, and they receive the funding for that work by the judgment of their peers. That peer review is sufficiently pure that it needs no external boundaries on it, because all scientists are interested only in what is the best question to be asked with the tools available now. And they will agree on that question when given the chance. That is the rhetoric.

The reality is that money talks. Peer review is one of the most risk-averse, socialist systems you have left in the United States. It is not corrupt. It is an absolutely honest, risk-averse selection system. So everybody knows that when you write a grant, you include as "to be done" some things you have already done, because you are certain they will work. If you have a really good idea, everybody knows that you first get your money for something that will work, and then you use some of that money on your idea, because it may not work. If you ask for something that may not work, you'll never get the money. Everybody knows that. Yet everybody pays lip service to another set of values, and the NIH grant forms require one to do so. In the "Aims" section they always ask, "How will what you do affect medicine?" So you answer as expected. Some people who are interested in viruses study AIDS because there is money in AIDS—not because they are really driven by the thought of a million babies' dying in Africa.

If you start with the rhetoric, all that religion can do is perturb peer review. If you take the reality, then what religion can do is make scientists more honorable and more honest about their own impulses and less risk averse in their work. No one I know would say from a religious perspective, "You can't do that experiment because it may offend my religion." But I would definitely say that you should not give money away, or take it, without answering the question, "What does your heart tell you that you are afraid of?" That is where the religion comes in. The separation of church and state is a great principle; it is why you and I can talk to each other. But there is no separation of church and state for the church of science. The church of science is a state church. I would just like to see the separation of church and state applied to the church of science.

What I'm beginning to understand is the role of your critique of science in your understanding of science and religion. My question made the assumption that science is an objective practice; hence for any subjective influences, such as religious influences, to affect it would be to corrupt it. And your answer is, You have misunderstood science, my friend.

Yes. Science is done by scientists; it's not done by God. Scientists have put themselves on the throne, saying, "I am the agent of absolute objectivity." They are lying to themselves and lying to you. They are, as we all are, wet weak mortals. They are made of tissues like us all. They have inside their skulls dark wet things that guide their thought. They are going to die like the rest of us. They should accept that and then do what they want with that gift. If they have the strength to say, "I accept all that but I'm going to run my lab," all well and good. But most of them have those fears and deny those fears for the simple matter that it gets them money.

Can I switch the emphasis a little bit and ask you, on a more experiential level, about the practice of religion and the practice of science? Some people say that these are two vastly different parts of their lives, whereas others speak of certain parallels in their experience of the world in these two areas. You've hinted that yours is more like the latter position, for instance, when you talked about the sense of revelation through a text and the sense of discovery in the study of the natural world. Are there other areas where, at the experiential level, you find parallels between these two?

Yes, I like that question a lot. In my tradition the three aspects of religious observance are prayer, the study of Jewish texts, and the doing of deeds of kindness, usually called good deeds. Kindness, though, is a specific concept, distinct from deeds of justice. Good deeds are also those beyond justice. So the doing of those three things is the sum of religious observance for a Jew today. Why? Not because that is the original intention of the covenant of Abraham, Isaac, and Jacob to their descendants, the Jews of today, but because that is what is left for us to do in the two thousand years since the Romans destroyed the temple at which we were to bring sacrifices.

These acts, then, stand in place of the sacrifices that God asked of our ancestors as our side of a covenant that assures our persistent survival and our capacity to redeem the world. So our deeds today are the deeds in exile, left to us now that we no longer have the place in which to work inside the covenant to bring the eventual redemption of the world. Your religion and mine come out of that moment. It's worth another conversation at another time [to explore] how we each learn that lesson of living in exile differently, because that temple, having been destroyed, was destroyed for both Christians and Jews. Therefore my tradition is the tradition in exile now. For me as a serious Jew, you are asking a very important question. The question, to my mind, is, "In what ways is the doing of science like the carrying out of sacrifices at the temple?"

That's a fascinating way to put it.

In what way am I to acknowledge that the very fact of life and death is in God's hands and do so every day? The temple was a slaughterhouse. You couldn't be a Jew and not know that the life you bore out in your herd was going to be killed on an altar to God. The firstborn was killed. Modern Western civilization says that's disgusting, yet it then carries out the slaughter of human beings at a rate that far surpasses any in previous historical time. What we have in my tradition is an acknowledgment, by the death of our animals' firstborn, that we are not in charge of life and death. The redemption of our firstborn sons is a tradition carried out by Jews today. For three thousand years, regardless of the temple, it is commanded that your firstborn son be given to temple worship. Not for sacrifice but be given to work in the temple, unless a priest redeems that son so you can have him back. So my grandfather redeemed me—and, God knows, one wants that, because it is holding on to an aspect of the original covenant.

Science is like sacrifice in that original covenant. You kill experimental animals to understand. You take dominance over the natural world for the sake of understanding on the one hand and for the sake of saying thanks on the other. But in both cases you act against the natural world for your own interests, your own survival, and your own covenant. That's where it is the same. What is left over, in the absence of the temple, is much more civilized, and its similarities may be harder to see.

I would also say that the doing of good science is the doing of a good deed. If your intention is to help sick people, then the doing of science is a religious act. But my first answer is the stronger answer. My religion, the way it was intended in the text, includes the action of animal sacrifice. The secular version of this is a laboratory that sacrifices animals to understand human disease.

I want to ask you about a change in the broader cultural climate that many people today are perceiving. In numerous ways the modern world has been understood through the opposition of science and religion. Today that opposition seems to be dropping, for example, through the work of centers like your own. There is a widespread sense that we are looking forward to a century in which humans will do great science but will no longer do it in the form of a battle with the religious side but, perhaps rather, through an alliance between the two. I wonder, is this sense of a major shift overblown, in your view? Is the prophet speaking too early? Or do you see that sort of change in the air?

Well, I don't see that change at Columbia. If I consider the 500 people in the faculty of the arts and sciences departments, 300 are tenured, and, of those, 150 will be scientists or social scientists. If 5 or 10- percent of those 150 tenured faculty are interested in what we are doing, that's a lot. If it ever flips from 5 percent to 50 percent, it would be a different institution. My work is to move it along in that direction, but I certainly am not coercive. We just make interesting things happen and see who comes.

Now the medical school is a different question. With the medical school you are probably dealing with a large number of people whose religious impulses are repressed for professional reasons. The liberation of physicians and medical scientists is an act that any religious organization can undertake, as it undertook it for patients in the hospice movement. There is an obligation not yet met: to provide for healthy doctors what is provided for terminally ill people. I'm being dramatic, but I really believe this. One doesn't need a hospice as a scientist or a doctor; one needs what a hospice provides: the certainty of safe human contact and a place in which to admit one's fears. If you have ever been sick and seen a doctor who doesn't like telling you that you are not well, you get the burden of that doctor's weakness. The institution needs to protect you against that. And for that, nothing beats a religious context.

There was a very important article in the *New England Journal of Medicine* by a colleague of mine in about July 2000. Richard Stone wrote an article arguing that doctors should not know about the religious observance of their patients, and doctors should not use religious observance in their practice because it instrumentalizes religious practice into an aspect of health care and thereby diminishes the religious content of what happens to zero. It is a very interesting ethical question. I think the most fruitful place to ask your question about integration would be in a medical context, not a basic science context. With a basic science context I don't think it is going to be the case that most people who pass through the selective system of tenure are going to be people who remember what it is to speak from the heart. The "either/or" is usually constructed as science or religion. I would say that the real "either/or" is emotional honesty or tenure. That has been my experience.

What is interesting about your answer is that you give the word religious the broadest sort of meaning, one that would incorporate most of the world's religious traditions. You come close to equating it with being fully human.

Correct. Initially, as I found myself falling into this position, I found myself drawn to it by an obligation to accept all the data, including emotional data. That's the point. When somebody says, "Scientists are different from religious people because they are totally objective," what I hear them saying is, "Scientists are less than full people because they are afraid of their emotions." That fear that I see drives a religious use of science, which is inappropriate for science and doesn't in the end work.

Conversely, if one begins with a deeper understanding of the human person and expands it outward—from physical reductionism, from biological reductionism, from reduction to brain states—expanding it to include the emotional, the moral, and the aesthetic realms, would not religion inevitably enter in?

Absolutely. I have a colleague at Columbia, Rita Sharon. She is a Harvard M.D. She has been an internist at Columbia Medical Center for twenty years. She runs a program in humanities and medicine at Columbia, and she has a writer-in-residence program to have doctors and medical students write about their experiences. Her work fosters these people to write about their religious experiences, to get them to write beyond the chart, and to say something about how they feel about a patient. Dr. Sharon was born Catholic, and I don't know her religious observance, but to me she is a religious person in a medical context.

What we have avoided are the particularities of one religion's observances against another. I have run into that issue over and over again. By the luck of my life, I have lots of friends of different religious backgrounds, and I'm much more comfortable sitting and talking to somebody who is a serious Catholic than an absolutely secular and therefore emotionally frightened Jew. I have addressed audiences after mass, and I have had no difficulty doing that, although there are points of religious ideology in my religion and between religions that are unbreachable. You must either walk away from a person in need or say we have to agree to disagree on that if we are to keep talking. What we share as religious people, I think, are two elements: curiosity and a humility about the limits of human intelligence. I want to know how far I can go, but I no longer claim I can go as far as I want to go.

That's a beautiful summation of many things you've said since the very beginning of the interview. The word humility conveys a transformation of character and a particular attitude about one's science and one's life in the world. It seems to express a deeply religious orientation.

Yes. One person I've met who understood this best is Sir John Templeton. A very misunderstood character, as I suspect that, on a small scale, I am a misunderstood character. I think most people, when they hear him speaking of the theology of humility, don't believe it; they think it's a cover for the theology of one version of Christianity. But I believe it explicitly, from meeting him and reading his writings. There are indeed behaviors like humility that are driven by emotional states common to all religions.

I think religion is an aspect of the human species. It emerges the way that family structures emerge. To deny it because it cannot be pinned down, or to deny it because it takes on different manifestations in different parts of the world, is like denying meaning because meaning is expressed in six thousand different languages. To say there is no religious truth because there are so many different religions is like saying there is no literature because literature is written in so many different languages. This is just not the case. The case is that different people are born with different languages, and they come to the same point. The poles of that point are humility on the one hand and curiosity on the other. Humility without curiosity means you are a zombie. And curiosity without humility means you can be a fascist—you can really kill somebody.

What has struck me in this conversation is the way that you have consistently and relentlessly challenged the dichotomies that are part of the popular perceptions of science and religion. Science as objective and religion as subjective? You've taken this dichotomy away. Religion as otherworldly? No: in your view religion is a humanizing influence on medical science. In discussing the differences between the religious traditions, you emphasize the richness of particular features of the Jewish tradition, and yet at the same time you find general features of the religious face of humanity that run across traditions. You seem in many ways like the great attacker of dichotomies.

I don't mind being that. I'm very impressed with your line of questioning; you are making me see myself clearly, and this is terrific. You'll notice that we left out one word, which I happen to be allergic to, and I'm very grateful you did not mention it. That is *spirituality*—the idea that somehow the religious impulse is unworldly or otherworldly. Spirituality defaults on the fact of curiosity about the world. But the religious impulse, at its core, draws on the obligation to take care of other people.

Conclusion

Philip Clayton

A number of standard typologies exist for conceiving the various relations between science and religion. The most famous is the four-part typology, developed by Ian Barbour in *Religion in an Age of Science* and numerous other publications, that emphasizes the contrasts between the models of conflict, separation, dialogue, and integration. (This typology is described in more detail in Jim Schaal's introduction.)

And yet, when one reads these twelve interviews with practicing scientists who are attempting at the same time to live lives of faith, one is struck by the inevitable inadequacy of all such typologies. Most, if not all, of Barbour's four famous categories make an appearance in *each* of the twelve stories. Some of the scientists, like Ursula Goodenough and Paula Tallal, start from their science and through it come to a deep sense of spirituality. Others, like Robert Pollack and Donna Auguste, start with a clearly formed religious identity. One can see how that identity serves as a touchstone for their lives in science. Still others, such as Jane Goodall and Thomas Odhiambo, begin from a pervasive sense of spirituality that colors all that they think and do. Each has occasionally experienced conflict between science and faith, and all understand their science as in some ways distinct from their ethical and spiritual beliefs. Yet each is involved in serious reflection and dialogue between matters of faith and matters of scientific knowledge, and each has succeeded in integrating these two disparate domains into the context of a single life.

Actual lives are not as neat as conceptual typologies. We all know this intuitively, of course. But the point is brought home in a particularly powerful way by the stories of individual lives and individual struggles toward reconciliation and integration. Such is the power of biography. One cannot walk away from the encounter with these remarkably articulate and reflective scientists without a renewed sense of the diversity of the human quest for meaning and excellence.

Nonetheless, despite the differences among the speakers, one commonality stands out clearly. The story of *each* of these twelve scientists undercuts a certain widely held stereotype concerning "science and religion." According to the stereotype, science demands complete neutrality and objectivity—an objectivity that can be achieved only by people who either have no religious commitment or are hostile to religion. Having read these twelve stories, one might be forgiven for wondering how that stereotype ever arose in the first place. After all, these twelve individuals clearly are deeply committed to the scientific project and to excellence in their particular scientific subdisciplines. All enjoy international reputations for the quality of their work. And yet in no case does one see signs that spiritual or religious commitments have detracted from the practice of science; indeed, in most cases the scientists argue that their faith has actually *increased* their effectiveness as scientists.

In their diverse ways these stories chronicle a natural partnership between practicing science and living faith. Khalil Chamcham's belief in God encourages him in his search to find and to formulate the mathematic order expressed in the fundamental laws of the universe. Ursula Goodenough's quest to understand the evolution of life and her sense of "the sacred depths of nature" are inseparable: the one perspective supports the other in a seamless vision that is as clearly scientific as it is spiritual and ethical. Indeed, one could even speculate that Pauline Rudd's understanding of the relatively large biomolecules that she studies is actually enhanced by the intuitive skills that she has developed in her spiritual practice. She has to think her way into the inside of these macromolecules, using numerical data as her only guide. Only when she can translate the rows of numbers into a mental picture of the undulating cell walls can she hypothesize where and how the proteins attach and then proceed to test these hypotheses in her lab. Such vision and intuition, one realizes, is honed through her spiritual practice. Likewise, it seems plausible that the clarity of thought that Hendrik Barendregt achieves through countless hours of demanding Buddhist meditation has helped him to make the theoretical breakthroughs needed to develop the lambda calculus. None of these four scientists confuses the spiritual and the scientific dimensions of their lives, yet all use the skills and insights from the one to help them gain a deeper understanding of the other.

This first point of connection, then, between practicing science and living faith concerns the *practice* of science and faith. A given individual engages in two different types of practice. Invariably, the one kind of practice will augment the other. This cross-fertilization is the first and perhaps the most intimate bond between faith and science for these twelve scientists.

The second point of connection, which is perhaps even more clear, concerns the dimension of values and ethics. The scientists being interviewed draw deeply upon their religious and spiritual beliefs as they struggle with how best to use the knowledge that is produced by their professional work and with how to disseminate the results of their work more effectively. Satoto found that teaching good nutrition principles alone was not sufficient; only when he learned to combine his scientific expertise with motivations drawn from the Qur'an was he able to do science that made a difference. The result of this partnership has been the implementation of nutrition programs that are credited with saving hundreds of thousands, if not millions, of children from malnutrition in Indonesia and other predominantly Muslim countries. Similarly, Paula Tallal found her spiritual values guiding her away from pure science and into the unfamiliar world of business. The company that she founded with Michael Merzenick, Fast ForWord, has helped thousands of children to overcome neurological deficits and to learn to read. Yet the entire effort was motivated not by scientific results alone but by a rich marriage of Tallal's science with the fundamental spiritual and ethical values that she holds as a person.

The partnership between science and faith hones practice, then, and it supplies the values that guide leading scientists in the application of their work for the betterment of humanity and the world. But, third, the partnership also yields theoretical fruits. As a physicist Khalil Chamcham struggles for means to express the fundamental intelligibility of the universe, and he finds inspiration for his work—but not a substitute for it—in the religious idea of a cosmic creator. Jane Goodall's groundbreaking role in helping to establish primatology as it is practiced today is inseparable from her sense of a fundamental spiritual connection with the chimpanzees that she studies. The links are not only about practices and methods, though they include these as well; they are also directly theoretical. Goodall broke with the Cartesian dualism that was dominant in ethological fieldwork (and at Cambridge University) when she began her research—a dualism according to which apes were closer to machines than to intentional human agents. Only when she challenged that theoretical paradigm was she able to make the discoveries for which she is now justly famous.

Before noting a fourth and final common feature that arises in these interviews, I should pause to note something that is *not* present in them. None of the discussions chronicled in these pages focuses on doctrinal disputes or on lines of proof or disproof that allegedly run between science and religion. None of these scientists has attempted to establish the truth of her or his religious beliefs based on scientific conclusions, nor has any challenged accepted scientific results on the authority of religious premises. The politically charged debates that fill the pages of American newspapers today, the heated debates spawned by boards of education in various states, are (thankfully) absent from these pages. None of these scientists claims scientific warrant for the view that order in the physical world can be traced back only to an intelligent designer; none questions the status of biological results based on worries about the metaphysics of Darwinism or concern that "evolution is only a theory"; and none challenges the results of the neurosciences from the standpoint of his or her belief in soul, mind, or God. (Some believe in the existence of souls, minds, and God, of course, but none finds that these beliefs undercut the pursuit of science.)

This is not to say that the types of issues just listed are unimportant from a philosophical or metaphysical perspective; indeed, they are among the most urgent questions that philosophy faces today. But the scientists interviewed here have greater clarity about the *status* of such questions than one usually finds in popular presentations. They realize that these are questions that *arise out of* contemporary scientific practice and its results; they are philosophical speculations that concern the *interpretations* and the *implications* of science rather than parts of science itself.

Indeed, this clarity may be one of the most important fruits of this collection, in which leading scientists have spoken openly about the boundary issues between science and religion from their own perspective. A close reading of the interviews yields a much clearer sense of what are, and are not, scientific questions. One begins to see three separate spheres, with important relations between them: the practice of science, the values that guide the implementation of scientific results, and the realm of philosophical and theological speculation. The danger, it turns out, is not that scientists sometimes hold religious beliefs. The real danger, it seems, arises in the work of those authors—often nonscientists—who *confuse* science and religion, seeing them as identical or using the arguments of the one to directly undercut (or to "prove") the core premises of the other.

All these considerations, when taken together, signal what is perhaps the greatest strength of the discussions that weave their way through these pages. The approaches taken by the scientists and the positions they defend demonstrate *an integration of the practice of science with the life of faith—an integration that points in the direction of harmony and mutual interaction rather than battle, dispute, and fear.* Though the scientists approach their topic in sometimes quite distinct ways, and although they answer in different voices, each is involved in an explicit quest for reconciliation. Reconciliation is an interesting notion. On the one hand, it presupposes that there is a *need* for reconciliation, which means that the two sides of the discussion are not, and never will be, identical. On the other hand, it presupposes that there is a *possibility* of reconciliation—that is, that the differences between the two sides are not permanent, essential, or unbridgeable.

In the end, paying tribute to this quest for personal integration may be the most distinctive and the most valuable contribution of this book. Many volumes seek to establish the compatibility of science and religion through careful theoretical arguments. But there is something particularly powerful in seeing the integration happen where the rubber actually meets the road: in the personal lives of scientists who talk openly and honestly about their dual identity as men and women of science and as people of faith. Somehow one gains a deeper and clearer understanding of the perennial issue that we call religion and science by observing the various ways in which practicing scientists actually achieve a harmonious working relationship between the two in their day-to-day lives. Of course, achieving reconciliation at the personal level does not replace the task of establishing a full integration at the conceptual level, nor does a personal reconciliation render the conceptual task moot. But success at the personal level does help to increase one's confidence that theoretical reconciliation is possible. In that sense this book may supplement the more theoretical presentations that are now widely available. After attaining a sense of the personal side of the so-called science-and-religion debate, students, professionals, and laypeople may find it easier to work their way into the more abstract theoretical disputes. And who knows: perhaps some of the suggestions made in passing by these twelve scientists will subsequently be developed, by them or by others, into full-fledged theories about the science-religion relationship and will play a role in the future evolution of the field.

At the end of the day, with the sounds of these twelve voices still ringing in one's ears, one may well wonder what was ever attractive about the idea of treating science and religion as sharply separate spheres in the first place. The natural movement toward reconciliation, the myriad ways in which science and faith interact in these individual lives, is hard to challenge. Clearly, these scientists have preserved their distinct identities as people of science and as people of faith—yet without freezing difference into dichotomy. They, along with the editors, hope that these models will serve to empower other scientists, and all those interested in science, to engage in their own individual quests for reconciliation. The answers to which different people come will not be identical—nor should they be. But the quest itself is nonetheless shared by many in our culture today. I trust that the answers obtained will be as rich as the quest is unavoidable.

Acknowledgments

This book owes its existence to a seven-year international research project and thus to a very large team of participants and administrators. The founder and director of the Center for Theology and the Natural Sciences, Robert John Russell, helped conceive the SSQ program and served as its senior adviser throughout. W. Mark Richardson (program director, 1996–98, and coinvestigator, 1999–2003) and Philip Clayton (interviewer and comoderator of the seminars, 1996–98, and principal investigator, 1999–2003) provided academic leadership. SSQ was administered by a top-level staff team—including Catherine Thompson, Anne Bade, Katherine Farhadian, Clare Olivares, Stephen Long, Sydney Prince, Bonnie Howe, Holly Vande Wall, Kirk Wegter-McNelly, Dale Loepp, and Shelly Dieterle—and was led during its last four years by Jim Schaal, who served as program director from 1999 to 2003.

We are grateful to the John Templeton Foundation and its founder, Sir John Templeton, for its generous financial support of the SSQ vision. Other direct funders included the American Association for the Advancement of Science, the Center for Theology and the Natural Sciences Science and Religion Course Program, the Episcopal Church Foundation and Trinity Institute, the Metanexus Institute on Religion and Science, the Procunier Family Trust, and the Tokyo Club. Among the nearly fifty program partners that worked with SSQ during its seven years of existence, we wish in particular to acknowledge collaborations with the United Nations Educational, Scientific, and Cultural Organization (UNESCO);

the International Society for Science and Religion; the Center for the Study of World Religions at Harvard University Divinity School; the Université Interdisciplinaire de Paris; the Romanian Academy; the Pontifical Council for Culture; the Université Al Akhawayn in Morocco; the National Academy of Sciences in Kazakhstan; and the National Institute for Advanced Studies in India.

Several people deserve appreciation specifically for their support in the preparation of this book. Philip Clayton conducted the majority of the interviews, and we wish to thank Gordy Slack for his interview of Henry Thompson, and Mark Richardson for his supplementary interviews with Thomas Odhiambo and Jane Goodall. Sarah Conning, Bonnie Howe, Jessica Hazlewood, Kevin Lucid, Catherine Thompson, and Holly Vande Wall helped with the transcription of the taped interviews. Holly Vande Wall and Anne Bade assisted with the administration of the book project. Zach Simpson worked as assistant editor during the final phase of the project to make sure that the interviews actually made it into print. We are grateful to Robin Smith, former senior executive editor for the sciences at Columbia; to series editor Robert Pollack; to the editorial staff at Columbia for their professional work on the book; and to Polly Kummel for her expertise as copy editor.

Above all, we wish to acknowledge and thank all the 123 scientists involved in Science and the Spiritual Quest. These men and women took time away from demanding careers to travel to Berkeley, Stanford, New York, and Paris, sometimes facing professional backlash for their willingness to publicly address religious issues. Not only did they write thoughtful original papers for the workshops but many subsequently spoke at SSQ public events around the world. In their participation they modeled curiosity in exploring issues of science and spirit, consideration in listening carefully to their colleagues, and courage in voicing their own doubts as well as their convictions. May their quests continue, joined by all those who seek insight in the dialogue of science and the spiritual quest.